Wolves and Honey

A HIDDEN HISTORY
OF THE NATURAL WORLD

SUSAN BRIND MORROW

A MARINER BOOK
Houghton Mifflin Company
BOSTON   NEW YORK

FIRST MARINER BOOKS EDITION 2006

Copyright © 2004 by Susan Brind Morrow

For information about permission to reproduce selections from
this book, write to Permissions, Houghton Mifflin Company,
215 Park Avenue South, New York, New York 10003.

Visit our Web site: www.houghtonmifflinbooks.com.

Library of Congress Cataloging-in-Publication Data

Morrow, Susan Brind.
  Wolves and honey : a hidden history of the natural world / Susan Brind Morrow.
    p. cm
    ISBN-13: 978-0-618-09856-9   ISBN-10: 0-618-09856-9
    ISBN-13: 978-0-618-61920-7 (pbk.)   ISBN-10: 0-618-61920-8 (pbk.)
1. New York (State) — History.   2. Natural history — New York (State).   3. Human
ecology — New York (State).   4. Finger Lakes Region (N.Y.) — History.   5. Natural
history — New York (State) — Finger Lakes Region.   6. Human ecology — New York
(State) — Finger Lakes Region.   7. New York (State) — Biography.   8. Trappers — New
York (State) — Biography.   9. Beekeepers — New York (State) — Biography.
10. Morrow, Susan Brind.   I. Title.

F119.M67 2004
974.7 — dc22   2004047278

Printed in the United States of America

MP 10 9 8 7 6 5 4 3 2 1

An earlier version of "Bees" appeared in *Harper's Magazine*.

Contents

first a bluebird's halo of blue from its fluttering wings

lukos= lux, light, as of eyes shining in the dark
lukospas: "torn by wolves," epithet of bees because they are generated from
corpses torn by wolves

— LIDDELL AND SCOTT
Greek-English Lexicon

The living appearance of a bird is seldom well expressed by dried skin, however perfect it may be, and in many instances a specimen gives no hint of the
grace and beauty furnished in life by the bright colors of perishable parts:
the eyes, the soft skin of bill and feet of many water birds, and, in rare cases,
the living feathers themselves. For there are some birds, notably the group of
black-headed gulls, some terns, and the larger mergansers, whose white
plumage is suffused at certain seasons with a most beautiful tinge of shell-
pink or rosy cream-color which is evanescent, and soon fades from the most
carefully prepared and cherished skin. The rich colors often found in the bill,
legs and feet also change with the process of drying, and it is a cause of surprise and regret to students to find how meager is the literature bearing
upon this important item of bird coloration . . . It early became apparent to
me that if such data were to be recorded it must be done from actual speci-
mens, painted, in short, from living or freshly taken birds, before the settling
of the bodily fluids or the disintegration or absorption of pigments could
take place. This, it may be said, is frequently a matter of only an instant.

— LOUIS AGASSIZ FUERTES
The Birds of New York State
Ithaca, N.Y., 1910

Everyone has noticed the influence of the American elm upon the abun-
dance of the Baltimore oriole.

— ELON HOWARD EATON
The Birds of New York State
Canandaigua, N.Y., 1908

Wolves and Honey

1

THE WOOD DUCK

LAST NIGHT I DREAMT I saw Bob Kime. I knew we were saying goodbye. I held him tight. Then he took off his jacket and gave it to me. It was a hunting jacket, soft and old, sort of bruised, I thought, and very dear. And then he was gone.

I always thought of Bob as my own particular friend, but at the funeral home on Friday people were lined up down the block, people I didn't know. We waited in line for an hour and a half just to get into the room to approach the open casket where his body lay. Shawn was standing beside the casket, having very much his father's face. Had I seen the picture in the back? he said. A photograph tacked on a board among dozens of others — of Bob with his dogs, with Shawn, with snow geese on the ground at their feet — and with them one of me and Bob in our bee suits in his old red pickup twenty years ago. Last Sunday I almost called him up to ask about a hive. But then I thought, Bob will think this is pathetic, my calling like this, as though nothing has changed after all these years. If only I had called him. For on Monday he shot himself.

Like the many times I have gone out to watch the moon rise, only to find it has risen, huge and gold and silent in a place where I have failed to look, I had missed the point, and the point was aimed deep into my own life, into the golden territory of the familiar.

At the funeral on Saturday morning Terry was there, sitting in the back row a few feet from where I stood. At first I didn't see him.

Terry is in his sixties now. His black hair is white. But there were the huge sloping shoulders, the same large head, the gold outline of the glasses he has worn these last ten years as he turned to laugh with the person beside him, some stranger on edge, as we all were, in the dim yellow light of the crowded room, Bob's soft profile, like something set in stone, occasionally visible through the rows of people shifting like rows of corn in the wind. When everyone rose to leave after the service was over I leaned forward and slipped my fingers into Terry's large rough hand. "Well, Suzy," he said, "all your buddies are gone now."

When I was growing up we thought Terry was a Cherokee Indian. It turned out that he was simply from California, and even though he had a crew cut and was something of a math whiz, and was also, it occurred to me only later, all the while a scientist and a chemistry professor at Cornell, he was our only real experience of the sixties, of an unconventional person. For a large man, who could easily have been threatening, he had an atmosphere of total ease, of kindness, and I had taken refuge in the safety of his presence for maybe thirty years.

Later Lan and I drove down East Lake Road where the Kime fields lay in soft shining squares of pale green oats and darker soy and golden wheat, patched like a lovely quilt in a rolling sweep down toward the dark blue line of Seneca Lake. The Kime barns and dwarf apple trees and farmhouse — large and white and square, the way the farmhouses are there, with a square windowed cupola on top where one can sit and see out over the fields — stood by the road lined with maple trees, as they have stood from the earliest days of my life.

Beside the bluestone marker just beyond — a gravestone carved in the shape of a dog, a curious antique — a dirt road leads down to Anne and Terry's cottage on a bluff above the lake, the burnt-out shell of an old log cabin of dark wood, polished now and screened, so that it recedes within a line of tall white pines and is almost invisible.

Anne has cancer, and has taken on a kind of translucence after these last months of illness, as though her fine blond hair were re-fined to silver. Her blue-green eyes had a radiance that surprised us as we walked in and saw her, for the first time in maybe a year.

We sat and watched the sun go down across the lake below through the broken black outlines of the trees. The faint flicker of a rainbow formed for an instant in the low sky to the north, as though it were the rim of something suddenly visible, a shining fragment of the rim of a halo. The last light fell in a wave of gold that swept quickly around the room, settling for a moment on each of us in turn. We sat quietly talking in the dark, in what seemed like a box of deep blue light, as we had in summers past, so that the evening had about it a sense of timelessness.

I reminded Terry of how once he said that everything operates on the level of four basic elements, their combining and breaking down, and that we are all "just some spectacular sideshow," as though all the desperate suffering of life were simply an elaboration of this basic principle.

"What is it that makes a human being?" he had said. "What defines being human? Falling in love. And what is that? Seeing something ordinary as . . . numinous." He thought a moment. "Seeing. The intensity of that focus, that concentration of energy, would be the heating up in which some significant transformation could take place."

Last Monday night a friend of mine called to say that she had heard a scream, a terrifying, almost human sound, and outside found a newborn fawn, still wet from its mother, and all around it black vultures in the trees.

"Bob talked a lot of people out of trees," Terry said, remembering how I first went to him, just wanting to be around that kind of man, a hunter, the year my brother died, "but nobody was there for him."

When we were children, barely able to walk, my parents would take us out into the middle of Seneca Lake and toss us off the side of their boat into the deep green water. Although we could float in our life jackets, and there was the electric touch of the water itself, the lake seemed dense and bottomless — heavy matter, like a skin not easily shaken free. We had an instinctive dread of what could drift up through that heavy medium from below — the immense primordial sturgeon, like pale ghosts, plated in hard ridges of leathery gray.

The lake was something that we knew by heart, through our bodily senses as they themselves were formed.

In those days there were only simple cottages in the bays, little clapboard houses of one story, painted blue or white or gray. The narrow water-worn docks of splintery wood stretched out into the water on thin pipes rarely more than twenty feet.

The fields behind them glittered with the multiplicity of summer life, speckled red beetles on the milkweed leaves, the fragrance of the milkweed unbearably sweet, its gummy milk bleeding into our hands, the seed pods, their skin like pale knobby velvet, pulled back to reveal a tight silver-white pattern of satin-rimmed scales. The seeds formed the body of a tiny fish — a fish made of silk you could pull to pieces and float away.

When we first came to the cottage it was full of old things: a kind of old pine green and teal blue tinged with gray, lined plates of pale blue glass, heavy stoneware, a fieldstone fireplace, and, before it, a bearskin rug smelling of bacon grease, and after we were there, mounted fish on the walls — the walleye I had caught in Algonquin Park that was patterned green and gold, with its tall reptilian dorsal fin (how often we would get the spines of fish fins stuck in our fingers in those days, and soak them out with Epsom salts).

My parents bought the place with all its contents, and there were a lot of old books, Gene Stratton-Porter's *A Girl of the Limberlost* — the story of a girl who put herself through school collecting rare moths in the swamps of Mackinaw — and *The Keeper of the Bees*, about a World War I veteran dying in a war hospital, who got up and staggered away, and found a garden on the sea filled with flowers in every shade of blue, a garden filled with skeps and bees.

World War I and, after the terrible shock of that war, the solace in the eternal presence of nature, were pervasive elements in the atmosphere of the place. My mother was formed by the aftermath of that war, and the books in the cottage were embedded with a sense of the time, like the musty smell embedded in their pages.

There was one green book, *The Bird Study Book,* with a golden moon pressed in relief on its cover, and flying across the golden moon a dark flock of geese. Years later the cover remained like a seal impression in my mind, although I had forgotten the book itself. One day in New York I called the astronomy department at Columbia University and said, "Can you see geese flying across the full moon?" Their reply, after I was put on hold for a minute, was "Yes. When there are geese flying across the full moon."

🦆

My brother David became a duck hunter in his early teens. We used to go out in the boat so he could practice sighting the birds in flight at a distance around the lake when the migrations came through in the fall. We were used to seeing flocks of ducks settled on the icy water near the crumbling old stone pier as our father drove us to school in the morning down Hamilton Street. They had a mottled quality that almost shone in the crisp clear air. Some were beautifully patched with white — buffleheads and goldeneyes among the canvasbacks and redheads.

One Christmas Eve David appeared on the porch in the dark in the moss green hunting jacket my mother had made for him by hand, with a brace of canvasbacks over his shoulder. My mother would later say, "How I remember his Adam's apple bobbing in his throat!" David, thin and blond as he was then, having recently come back as an eagle scout from Philmont, which made him even more of an outdoorsman — always up at 4:00. There he stood with the glove-soft white breasts of the ducks, their burgundy, oddly shaped heads spilling down the front of his jacket. How cold it was, the film of shining dark ice on the walk, the hard snow sparkling white beneath the trees, and my mother saying, "Well, you can pluck them outside!"

But David and I went down to the basement and spread out newspapers on the floor. I remember the sense of the gathered tension of the feathers as they ripped out from the skin with a soft puckering sound, the feathers coming loose in my hands, the soft inner down full of mites. Redheads, buffleheads, canvasbacks — the meat gamy and tough, tasting of fish, full of shot, the shot falling loose on the plate as you cut the dark-stained meat. The circular burn around the shot burned into the flesh remained, although we cooked the birds in wine for a long time.

When we were children David and I used to catch things just to look at them, and sometimes kill them to see what was inside. One summer we found a mudpuppy under the dock, purple and splotched, with gills that blossomed out like the purple buds of a Judas tree, and perfectly fingered hands. We buried it on the shore and later dug it up to see its beautifully articulated thin white bones.

My father was a lawyer, and we lived in town. But somehow for us as children our great experiences had to do with being outside. I have a photograph of David and me standing in the Canada woods — David in a soft blue cloth jacket with a white blond crew cut, me in faded corduroy lined with plaid. We are tiny beneath the tall trees amid the masses of green ferns. I am holding a magnifying glass toward the ground, and looking up. Thus is a life spun together through layers of sense impressions, the light speckling through the trees, the smell of dead leaves and damp earth. For me the elusive thing of value has ever been the golden light of kerosene lamps, walls of thin blond wood, tarpaper tacked over a table, some smell of damp, and just beyond — the rich outlying darkness.

When David died in 1981 I was studying Greek in New York. I still have taped above my desk a fragment from Ibycus:

Tou men petaloisin ep' akrotatois
Izanoisi poikilai aiolodeiroi
Panelopes lathiporphurides te kai
Alkuoves tanusipteroi

In these lines of early Greek poetry key words are mysteries, because the author made them up. And they were never used again. All one can do is break them down into their component parts, and then guess what the composite might mean. It reminded me of oolitic stone: in the words, as in the thing described, the beauty lay in the flaws themselves, the irregularities — the speckling, the splotching, the mixing up.

There was *aiolodeiroi* — throats that shone with their dappling of color — with *aiolos* implying a moving brightness, a glittering, a speckling, as in *aiola nux*, the starry sky.

The fragment went something like this:

Among the highest leaves they sat —
The mottled ducks, with throats
That almost shone;
And halcyons
that secretly grow red
with wings outstretched.

One could only think, reading this, of the American wood duck with its shining splotches of color, its white speckled throat, its silver-blue wings like the panes in cathedral windows. I don't know if there is another duck that lives in trees. The wood duck was a rare bird when I was growing up. Its populations had been decimated by the nineteenth-century fashion industry. I had never seen one, only in pictures in books.

The hardest word was *lathiporphurides* — with porphyry, a word that means brightness itself, an emphatic doubling of the word for living brightness — *pur*, fire, the moving brightness of burning red, or the heaving of the sea with its glittering changing light. Here attached to *lathi*, meaning "in stealth." The Greek dictionary made a leap into the violence implicit in the color red, and translated the word "feeds in the dark."

I can't remember the day I met Bob Kime, he came into my life so quietly, and was so utterly familiar.

My father and I would sometimes stop by his house near the lake on Sunday afternoons when some of his friends were over shooting clay pigeons. The men would be standing in a line, with great seriousness of purpose, aiming and shooting down the little clay discs as they were flung into the air out of the machine with a rapid clicking noise.

I was never much of a shot, but when I was growing up it was considered important to know how to handle a gun. I had been target shooting from the age of eight. As a teenager I had my own Remington, and later even a pistol permit. There was a great deal of pleasure in sighting the discs as they fell rapidly through the sky, pulling the trigger, and seeing them shatter into pieces.

Bob would be standing in the line all the while, casually joking as we all were. When one of us missed, he would stop midsentence, raise his shotgun to his shoulder with a certain ease, and pick off the disc before it hit the ground.

He was an ordinary man of medium build, with dark hair and dark eyebrows. But he had a kind of antique face: soft features, eyes set a little wide apart, the kind of face one might imagine an American farmer having had a century or two ago — and indeed his family had been farming the land on the east side of Seneca Lake for a long time.

But most characteristic of him (so that one might not notice other things — I can't remember what he wore) was a kind of brightness. He had, one might have said, a beautiful radiance: he was a man who saw things, who saw things and understood them.

One October evening after we were friends he took me out to the Junius swamps. We stood in waist-high waders in the cold murky water amid the water-rotted trees, some still standing, with the faint pink hands of remnant leaves floating on frail elongated stems up to the surface, some gnawed down by beaver into flaking points like palisades. The sky was silver blue with a film of cloud, but we could see the stars come through in the early dark.

We watched the ducks fly up in gathered bursts, and tried to see what they were in the half-light, by the pattern of their wing beats, their patches of white. For some reason we didn't bother to shoot at anything.

At Christmas that year Bob brought me a wood duck. I had asked him that October night if he had ever seen one.

2

THE TREE
OF LIGHT

THE IROQUOIS SPOKE of a tree of light, on whose branches were
blossoming stars, and of a tree that all year long bore flowers of delicious fragrance and fruit of delicious taste — a concentrated vision
of the apple tree, whose white blossoms seem to be made of light as
they scatter like snow on the ground.

When George Washington sent his general John Sullivan to break
the back of the Iroquois confederacy in the summer of 1779, Sullivan
went on a scorched earth campaign. His intention was to burn every
trace of the militant branch of the Iroquois, the Seneca nation, to the
ground in retaliation for the Indian and British massacre of settlers at
Wyoming, Pennsylvania — a military action that had been planned
and carried out from the Seneca stronghold, Kanadesaga, on the
present-day site of Geneva, New York.

Sullivan's men thought they were going to a wilderness filled
with primitive savages, the Northwest Frontier, as the Finger Lakes
region was then called. They marched up through the Chemung Valley and north along the east side of Seneca Lake. Their journals are
filled with a sense of astonishment at what they found — not a dense
and terrifying forest, but open country, with orchards and cultivated
fields around the Seneca villages, which they called castles, of bark.

As Arthur Parker, the Seneca who became the first ethnologist at
the New York State Museum in Albany, pointed out in his 1913 monograph on corn, the Iroquois were such an agricultural people that

the Europeans attacked their fields instead of the Indians themselves. Parker quotes Sullivan's letter to John Jay, written on September 30, 1779:

> Colonel Butler destroyed in the Cayuga country . . . two hundred acres of excellent corn with a number of orchards, one of which had in it 1500 trees . . . the quantity of corn destroyed at a moderate computation, must amount to 160,000 bushels. I flatter myself that the orders with which I was entrusted are fully executed . . . We have not left a single settlement or field in the country . . .

Sullivan's men called one place "apple town" as they marched north along Seneca Lake, cutting the trees down. The apple trees themselves were like a buffering force, they complained, for the trees were thick-bodied and old, and the hard work of destroying them slowed down the expedition.

The entire ripe harvest was consumed in fire. In less than two weeks Sullivan's army reduced the Seneca to scavengers, wandering and hungry in the Niagara region, a tentative nomadic people. As Sullivan came north they vanished into the trees, offering no human resistance, though they took with them two of Sullivan's unwary officers, Parker and Boyd, and slowly hacked them to death. In his report to Congress Sullivan describes the shock of finding their headless mutilated bodies and then their heads — from which everything that could be cut away was cut away.

Sullivan's men made note of the hidden beauty of the place — a glacial landscape with rich deposits of soil set in long low hills rolling like parallel waves within a row of deep blue lakes. The cool deepness of the water tempered the climate and created the best fruit-growing land that they had ever seen. The name Geneva may have come to mind, but it had long since been coined by the French trappers who camped in shacks on the edge of the malarial swamps at the foot of Seneca Lake. It was part of a long established network of fur-trading outposts in the lake country, the critical gateway to the rich trade in furs across the continent, the American Silk Road.

In 1787 a beautiful woman from New England sent scouts to the Northwest Frontier with the intention of setting up a colony in the wilderness. Jemima Wilkinson was near death in a fever when a vision came to her of angels descending from the sky saying "Room, room, room. There is yet room, and all may drink from the water of life, which is free to all without price." Wilkinson said that a spirit of light had come upon her and scorched the human life away, leaving her neither male nor female, but merely the vessel in which the luminous spirit dwelt.

Wilkinson embodied the Quaker principle stated by Pythagoras: that friendship is true equality. In the landscape of the Revolutionary War she became the Publick Universal Friend. Wilkinson's contemporaries, the founding fathers of the United States, saw themselves as equals, with the common goals and qualities of male landowners of European descent, with everyone else subordinate to them. But Wilkinson received all people. Among her many disciples were men and women, rich and poor, black and white, colonists and British soldiers alike. She is described as having had healing powers that bordered on the miraculous, and traveled widely in New England to use them, visiting the wounded and dying on both sides of the battle lines. She referred to death simply, without fear, as "leaving time."

Wilkinson's enemies said that she had the audacity to present herself as the second manifestation of Christ, in the body of a woman. She was stoned by an angry mob in the urban northeast. People wrote of her long black hair flowing eccentrically, undressed, down her back, of the eerie contrast of her lovely face and strange male voice, of her eyes, which were startling and intense. Some compared her to Scylla, a monstrous female trap for trusting innocents.

At the age of forty-four, Wilkinson, on a black horse with a blue velvet saddle, rode up the east side of Seneca Lake to found a New Jerusalem in the American wilderness. She followed the route Sullivan's army had taken ten years before, traveling north from the swamps of Elmira, where she and her followers were knee deep in March snow and mud, past the spectacular waterfalls along the lake's southern shore. One night they heard the roar of the falls from the Keuka Creek across the water, and rode around the lake toward the sound, past the sad remains of Kanadesaga, and down the west side to Kashong, where the French trappers there told them stories of the richness of the place and showed them remnants of the apple trees that Sullivan's army had failed to destroy, the hacked stumps sprouting thin green clusters of twigs sealed with tight March buds.

I could think of the town of Geneva, New York, where I was born, as a landscape with figures drifting over it like ghosts: Sullivan's men making their way north along Castle Creek to Kanadesaga, or myself walking up the huge slabs of tilted blue slate that made up the sidewalks on Castle Street where the ruins of Kanadesaga had long since become the New York State Agricultural Experiment Station. I might be fourteen or fifteen, in the freshwater light of the late winter thaw, the sound of the swollen creek rushing below, on my way to Anne and Terry's house above the creek with something I had proudly learned to bake, butter pastry or a thick perfect round of Scotch shortbread wrapped up in wax paper and string, on the day they gave me a book of woodland watercolors with the inscription, "May your life be filled with the dances of bees . . ."

I could think of the place as a kind of archaeological layering, the top layer a mix of local elements with the homogeneous overlay of contemporary American life — the concrete blocks of malls on the high terraced land above the lake where Hamilton Street becomes Route 5 and 20, where the inn named after Lafayette, who once spent the night there, burned down and became Wal-Mart or Kmart or one of the interchangeable chains that sell fried beef or dough or fish; or the sand-colored mountain of landfill on Route 414 where white gulls swirl above slow-moving diesel trucks that hour after hour, under clouds of black exhaust, haul truckloads of trash up from New York three hundred miles away. Yet through the layers would drift a distinctive quality like a flavor or a smell signaling something indelible, a native atmosphere.

The essence of the atmosphere, like its weather, lay in the presence of water. One could paint it with water words — *otter, water, winter, wet* — the root core echoing throughout like the cries of geese that marked the swing of the seasons. When we used to hear them, even in the house, in spring and fall, my mother and I would drive down to the lake where the shoreline was rimmed with storm-battered willows, to watch the trails of geese dissolve in thick black flakes falling all across the sky.

The other day she and I stood in Jerusalem Township on a bluff above Keuka Lake where Jemima Wilkinson's house still stands, "When there was suddenly a rushing noise," as Warren Smith wrote of standing in the same place in October 1959, "and, before we realized what had happened, a whole flock of wild Canada geese had passed directly over us, flying low, to land on a nearby field. If a visitation of angels from on high had come to bless us, Margaret couldn't have been happier."

Margaret Hutchins and Warren Smith often came to this place to paint, and I thought, as I stood with my mother there, how extraordinary the light was — a kind of watercolor light that seemed to rise in thick diaphanous columns from the lake itself. This must have been why Jemima Wilkinson had chosen to build her house here, I thought, in this peculiar atmosphere of otherworldly light.

I never saw the house where Margaret Hutchins lived near the lake below. It was torn down after she died of cancer in the early 1960s, but I often used to visit her grandfather's clock above the boat pond in Central Park in New York. I did not know at the time that it was a clock, or that it was a monument to her grandfather, Waldo Hutchins, one of the founders of Central Park and the Park's first commissioner. I was drawn to it by the words carved in large letters on what seemed to be the back of a long curving stone bench. The bench was gray and dirty then, soaked with urine smells, and not a place where many people would want to sit, though sometimes a muttering homeless man was there. I never minded dirt much and went over to the bench whenever I walked by to try to figure out the letters, which were in Latin, and so (to me as to most of the other people in the park) essentially in code: ALTERI VIVAS OPORTET SI VIS TIBI VIVERE, *If you can live for yourself you should live for others.*

The bench was on a discolored stone dais that was oddly etched with curving lines. It was years before I realized that it was the large circular back of a human-sized sundial, and that the lines incised below marked the precise progression of the bench's shadow on the vernal and on the autumnal equinox. A complex circular gadget rose from the center of its back. The gadget held the blackened bronze statue of a tiny woman in a wind-blown gown. Behind her were words, so small I almost did not notice them, NE DIRVATUR FUGA TEMPORUM, *May it not be destroyed by the flight of time.* As a teenager I was surprised to see the name Waldo Hutchins carved on the bench. I knew Waldo Hutchins in Geneva as Margaret's brother.

There I had learned time as a reliable circling of snow and thaw and low gray skies. In summer we cooked things that belonged to a precise progression of weeks — black bass caught casting in our bay, corn from the roadside stands in August, asparagus that appeared in the local market late in June, peeled a bright lime green to eat with Hollandaise spun white in the blender.

There was a week when the town of Geneva was washed with purple, when the lilacs were out. The color seemed to come on rapidly after the late March thaw. One such week years ago I stood in the doorway of the old Italianate red brick mansion that stood back away from Castle Street behind a dark screen of large unusual shade trees, one of the old houses that rose like an outcropping of odd rock from the rest of the town that had settled in streets and blocks around it.

Warren Smith, who was spindly and old then and bent almost double, stood beside me in the doorway with an odd formality, even courtliness, a threadbare fineness that was present in his long thin face, his fine gray wisps of hair. A tree beside the door blossomed purple in a ghostly film over long slender branches of silvery gray. I had never seen such a beautiful tree, and I asked him what it was. His voice was high and cracked like a scratchy old musical instrument as he broke into a small sweet grin. "Now that is an American tree," he said, and he told me about the redbud, how it came to be called the Judas tree, how its radiant blossoms, shining as they would once have done through the American forest in one of the loveliest colors in all the world, could have been seen as the blood of Judas, the blood of betrayal.

We had met that afternoon to have tea, and we went for a walk in his garden, following a circular path hidden by cut-leaf beech with their odd translucent tiers of light green, sharp, serrated leaves. The tower of the house, steep fish-scale slate rimmed with trellised iron, rose up darkly behind us. Warren Smith spent much of his life as a scholar in residence at Yale, where he edited the Walpole papers. But in Geneva he wrote small lovely books about his old Geneva friends, which he illustrated himself and had printed at the local Humphrey Press on Pultney Street. He was a very good watercolorist and filled the books with pictures and etchings. The characteristic Genevan, he wrote, was a spinster, an aged childless person like Margaret Hutchins or himself, around whom a house had formed as a living presence made of old things, the family relics — paintings, teacups, carpets, books — all priceless, private, and crumbling.

In his book about his friendship with Margaret Hutchins and their decades of painting expeditions, in which they often went about in canoes to paint the lakes, their cliffs and odd light effects, Warren Smith told how Dick Manzelman, our minister at the North Presbyterian Church, suggested in the 1960s that the town organize an exhibit of paintings by Arthur Dove, the first American abstract painter.

Dove's father was a stonemason and had built the North Presbyterian Church. In my life his elderly brother still lived in the old Dove house on the street above the north end of the lake. Arthur Dove kept a studio in Geneva during the Depression. Warren Smith quotes the New York critics describing Dove's distant home, not as Smith saw it himself, as a rare landscape of remarkable old trees and extraordinary light, but as a decaying backwater with rotting wharves and grass growing up through the broken sidewalks.

What the town was like without its grid of sidewalks we did not even attempt to imagine, though in 1877 the Luminist painter Worthington Whittredge painted a fair representation of the house that stood outside my childhood window with its high-columned porch and triangular pediment white against the sharp blue sky. It was his wife's family home. Whittredge grew up hunting and trapping on the Ohio River. He went to Europe to study painting for a decade, but was restless and dissatisfied with the formulaic academic and religious subjects he found there. He left abruptly one morning in August 1859 and made his way back to America, where, as he later wrote, tears came to his eyes when he saw a study of rotting logs in the dappled green light of a forest, a painting by Asher Durand. Whittredge realized that he had been searching for a teacher he already knew, nature.

In 1867 Whittredge traveled to Geneva to marry a woman he knew primarily through her letters to him. He painted that year a branch of the streaked and golden fruit of an apple tree.

⚓

Western New York used to be called the burned-over district, where the fires of the spirit, like forest fires, so burned people that they became harsh: wary and cynical. The remoteness of the region made it a draw for religious eccentrics like Wilkinson, and the moral reformers and con artists who came after her. Yet the burning-over created a sort of radiant ground for the clarification of key ideas and experiments in the development of American life, a hectic mix of authenticity and fraud.

The ground itself was thought to contain some kind of secret buried history. This began with what was known. The first settlers to farm the land around Kanadesaga agreed not to disturb the burial mound of raw earth, five feet high and forty feet around, that stood in an open field where a Mrs. Campbell, a captive from the massacre at Cherry Valley who was still alive among them, had witnessed one of the last of the Seneca tribal rituals: the sacrifice of a white dog at the end of winter.

Every year as leaves began to fall small groups of Seneca drifted in and sat silently beside the piled earth that contained the bones of their ancestors. As the shell of human life crumbled into the soil, and (as the Seneca elders said) passed into the roots of trees, their visits dwindled away, and after fifty years the ground around Kanadesaga became ordinary farmland.

At the White Springs Farm a mile to the east of Kanadesaga, the grading of a field unearthed a cache of unusually large human skeletons, suggesting a race of giants. The bodies were buried in the Indian way — in sitting posture, so that first the rounded surface of innumerable skulls appeared like pale melons emerging from the raw earth. So many bodies were uncovered that they were hauled away in carts, as though an entire village had died and been buried at once. The workmen were afraid to touch the bones with their hands, for they feared that the bones were infected with smallpox.

A cigar maker from New York named George Hull later hired a stonemason to create a crude giant representation of himself. He rubbed the statue with acid, gouged it with nails, and buried it on a friend's farm in Cardiff, east of Syracuse. When a crew of well diggers were hired to put in a well on the land, people from miles around paid to see the human fossil they unearthed, a testament to the holy words: There were giants in the earth in those days.

In September of 1821 a young farm laborer named Joseph Smith found golden tablets buried in the drumlin country a few miles north of Canandaigua, near the town of Palmyra. A guiding spirit enabled Smith to read the treasure text that lay close beneath the surface of the low-spun glacial hills.

Although the authority of Smith's revelation lay in the realm of the unsolved mysteries of antiquity — the lost tribes of Israel, the hieroglyphic alphabet with its spooky symbols: the severed head, the pervasive eye — the substance of the revelation was recognizably local.

As *The Encyclopedia of Religion and Ethics* recounts in 1951, "Having dug for fabled treasures among Indian mounds on the western frontier, Joseph Smith found a peek-stone . . . whereby 'Joseph the Seer translated the reformed Egyptian of the plates of Nephi' . . . The Nephites in their actions were the modern Redmen in disguise; in their mental habits they more closely resembled the local sectarians. Thus the speech of Nephi contains quotations from [a contemporary] *Confessions of Faith* and the speech of Lehi the heretical tenets charged against the Presbytery of Geneva, New York, in whose bounds Joseph himself lived."

In 1848 two young girls heard noises in the night in their family's rented house in Hydesville, a small farming village between Palmyra and Lyons. The noises, they said, were like the noise a hand makes rapping on the wall, the hand of a body buried in the basement. The ghost, whom they called Splitfoot, rapped out the answers to their questions in code. Quaker abolitionists in Rochester, believing that spirit manifestations were part of the natural human contact with the divine, heard about Kate and Margaret Fox and put the sisters on public display in a lecture hall in the dark. Strange knocking sounds were heard in the room. The word went out that in the Finger Lakes region there were "mediums," natural conductors to the spirit world.

The spiritualist movement, which the Fox sisters unwittingly initiated in upstate New York, coincided with the popular perception of the discovery of electricity. The ritual of the séance, described by

Ann Braude in *Radical Spirits,* ensued as a sort of contemporary parody of an electrical schema, with a dozen men (natural positive charge) and women (negative) seated alternately around a table in the dark, creating a circuit. The ceremonies took their cue from the followers of the Viennese physician Franz Mesmer, who a hundred years before had investigated what he called animal magnetism, electrical currents that run through living things, and concluded that everything exists within a glittering field of magnetic electricity. Spirits were instinctively felt to manifest in much the same way that electrical currents traveling through conductors mysteriously created light.

In the 1850s thousands of people (usually women, with their potently receptive negativity) discovered that they too could be mediums, and conducted spiritualist séances across the country, claiming to be able to draw back the dead in a theatrical whirl of clatter and flashing lights. Such ceremonies won converts throughout the scattered settlements of the West, in the concentrated intellectual circles of Boston and New York, and even in the White House, where Mary Todd Lincoln frantically tried to bring back her dead son.

By the time the Fox sisters confessed that Splitfoot was a joke, that they made the sounds themselves by cracking the bones in the knuckles of their feet, the spiritualist movement they had founded had spread around the world, giving rise to Madame Blavatsky's theosophy and the Society of the Golden Dawn.

In the autumn of 1848 the Smithsonian Institution in Washington sent a group of men to the Finger Lakes region to report on what was left of the Seneca ruins. E. G. Squier wrote back that on the Castle Creek at Kanadesaga, where the grass was beautifully green, the apple trees of the Seneca Indians still grew. Though they had been girdled or cut down by Sullivan's army, shoots had grown up from their roots and turned back into trees.

3

HECTOR

THE ROAD SOUTH along the east side of Seneca Lake traces the
path of Sullivan's march in reverse, through fields of corn and wheat,
under the old black Lehigh Valley bridge, past our cottage row, and
then the dark government signs warning the curious away from the
officers' quarters at the nuclear arms depot. A man I know lived in
one of these houses when I was growing up, when his father worked
as a munitions expert at the depot, where neutron bombs were secretly buried. Like his ancestor Joseph Smith, Gene Smith went on to
bring treasure texts into the light of the world. He became a scholar
of obscure languages and ultimately the Library of Congress repre-
sentative in Asia. Smith was later credited with rescuing what has
been called the largest body of philosophical writing on earth: thousands of manuscripts smuggled under the ragged clothes of refugees
escaping over the mountains into India from the monasteries of Ti-
bet, where the books were being burned. Smith arranged for the Indian government to trade them to the United States in exchange for
shipments of surplus American grain.

As children we knew the depot chiefly for its imprisoned herd of
white deer. Behind the high mesh fence along the highway we would
see the strange deer slip out from between the trees, a vivid white in
the soft blue light of evening. Scott Sampson, who wrote a hunting
and fishing column for the local paper acted as game warden for the

depot. Sampson once told me that the whiteness bred out of the streak in the tail. When the area was fenced in by the military in the 1940s a gradual mottling occurred over the yearly generations of trapped deer, until they were born white. This is what Sampson said, but we secretly believed that what was in the ground had altered them.

The state mental institution just south of the depot at Willard spooked us far more than the neutron bombs. We sometimes went to the town hall there for public hearings on the Indian land claims. The Cayuga were trying to get back the Finger Lakes National Forest, and, just north of Willard, Sampson State Park. The Indians would not have to follow state fishing regulations and everyone feared they would use gill nets and fish out the lake. A naval barge stationed offshore did depth tests on the elusive bottom, which was seven hundred feet down through layers of mud and sludge, making Seneca one of the deepest lakes in the world.

The road bends away from the shoreline at Willard along a cornfield sloping up to the town of Ovid and the roadside bar, the Golden Buck, where, in 1985, a girl wandered out into the night and was found the next day, stripped and dragged through the corn, her throat slit from ear to ear.

Farther south the land rises and becomes the Hector Land Use Area and the Finger Lakes National Forest, where once my father bought fifty acres as a kind of hunting camp on the Searsburg Road, the road that runs over the high escarpment tilting up toward the south between Seneca and Cayuga Lakes. The narrow lakes, a dense slate blue, lie embedded in the deep downward folds of the land, which rises up in the far distance on either side in a bluish-gold patchwork of terraced vineyards and sloping fields. It was, as my father liked to say, "God's country" — a primitive place. Not many people lived there. Some who did lived in little tin-roofed houses made of doors.

On our land a mud path led through tall white pines, deliciously fragrant in the sun, to a shack of weathered boards in a wooded dip beside a creek. A dozen hives belonging to a migrant beekeeper who came through once a year were hidden back away through the trees. The place was so quiet that the sound of insects buzzing in the sun seemed to belong to the light itself, and I felt, walking in, the rare and wonderful sense of being utterly alone.

Hector was far enough away from people to see wild things. On the dry stony side of the creekbed in spring were efts the color of cinnabar flecked with gold, their eyed skin as soft as wet sand. The solitary stalk of a cardinal flower, its rare deep red the true scarlet of nature, would stand in the fresh green light of the wet wood like a hidden flag.

In such a place one was always concerned with what was hidden in the dense layered coloring of the forest, like a woodcock in a pile of dead leaves. We used to go to visit the bird carver, in his house near where the landfill is now. We had heard he had almost been beaten to death as an undercover narcotics agent for the state police. When he was forced to retire on disability, his wife suggested he take up carving birds, sensing that for a hunter who had watched birds closely around the lake all his life carving would be an absorption much like hunting — a mental focus that would take him beyond restlessness and physical pain.

The first time we went to see him we asked if he had a blue heron. My mother had an affinity for this particular bird. We were always watching for herons on August nights when the sawbellies jumping for flies made liquid arcs of silver all across our shallow bay. The herons would fly by two by two just after the sun went down, moving low over the water like gigantic gray-winged bats in the dense orange light.

The bird carver had a heron, but he and his wife had hidden it away in a trailer outside. It was almost as though they were startled that the carving had come out so well — an original yet completely realistic representation that could only come from long familiarity and a clear individual perception of the bird itself. There was something precious about it, as there sometimes is about small things. The heron stood on its mirror patch among reeds they had made from wood peels and dried grass. I asked if he would make me a woodcock, and he did, hiding it as it is hidden in nature in a pile of leaves, looking like the leaves themselves, though in miniature the bird had a jewel-like quality, faintly tinged as though shadowed over with blue — the kind of bird only a local hunter would know where to find.

My parents got a letter last summer from a man who wrote, "You probably don't remember me, but I knew your son David twenty-five years ago. I am writing this at 4:00 in the morning. I just woke up from a dream. In the dream I saw David, still young, and said to my wife, 'That's the guy I told you about from Philmont.' It just made me so sad."

The man told the story of how one day, when their scout troop was camping in the mountains of New Mexico, they were being taught how to fly-fish. None of them could catch anything, though they tried for much of the day. When they were resting in their tents in the late afternoon he heard someone shout. My brother David, who had brought his own tackle, was pulling trout out of the stream every time he threw in his line. David said, pointing to the place in the stream where he was casting, "This is the kind of place fish like."

The man wrote he'd never seen anything like it — a boy who was just like the rest of them, but saw something in the landscape that they didn't see.

This was the kind of hidden thing one came to Hector to find. For years I thought I would make the place my home, and gradually assembled a heap of supplies in the shack — a kerosene lamp, a shovel, matches, an ax. I kept a little black notebook there for over a decade.

A deer path on the edge of the woods led to a small lake where aspen trees, the pale silver green of lichen, stood in a line on a narrow tongue of land in the water, the white undersides of their leaves tipping up in the wind. One night I brought my sleeping bag down to the lake near the mass of cattails on the shore. The ground was damp and cold, and I lay awake with a sense of the impenetrable hardness of the white, cold stars. In the middle of the night, a pounding vibration rose in the earth beneath me in the dark. I lay rigid as I felt it grow steadily stronger through my skin. As the running animal neared the place where I lay, it stopped and pawed the ground, then galloped away. In the morning I saw its track: a large buck on its way to the water to drink.

In Hector I sometimes saw a dozen or more crows settle on the branches of a tree in the forest, surrounding some large light-feathered bird. The crows would call out sharply in a circle, as though the sharpness of their voices in chorus could hurt the solitary hawk or owl, which sat silent and still in their midst. It was the mirror image of a line I knew, and it delighted me to think that what I saw was a common sight even in the fifth century B.C., for the line was from Pindar:

> Sophos o polla eidos phua
> Mathontes de labroi pagglossia
> Korakes os, akranta garueton
> Dios pros ornika theon.

I liked to translate the Greek:

> The one who knows is one who knows much in his own
>   nature,
> Those who learn are like crows, squawking at an eagle,
> The silent bird of God

but it wasn't especially accurate, for in Greek, as usual, the words meant more than one thing. *Sophos,* for example, later comes to mean philosopher, but Pindar was using it here of himself, in its original meaning of poet. Pindar was a professional poet, working on commission, but what he was after here was the archaic sense of someone who is privy to some mystery or secret and can crystallize that secret into words — poet in the sense of its origin in the Greek *poiew,* to make, to make out of thin air a mirror, a name to catch the soul of a living thing.

Sufi is commonly thought to come from *suf,* the Arabic word for wool, for the Sufis were essentially a wandering monastic order and like the early monks wore wool as a constant torment against their naked skin. But Sufi can also be traced to *sophos* in this original meaning of one who has a mystical inner knowledge. In Pindar's terms this knowledge was of nature, the living world. The *sophos* was "a man who with a bird finds his way to all his instincts / he hath heard the Lion's roaring, and can tell / what his horny throat expresseth and to him the tiger's yell / Comes articulate and presseth / On his ear like mother-tongue."

Eidos could be translated as knowing, but it is the participle for the primary verb "to see." "The sophos is the one who sees" is the literal meaning, who knows how to see, while the crows, Pindar's rivals and critics, can only hope to learn the imitative superficial knowledge suggested by the word *mathontes,* math, and hence are of one plain common voice, their coarse unmusical cries clumsily going after a mystery that remains to them forever silent, like the large light-feathered bird hidden in the forest tree.

The empathy of crows was a remarkable thing to see. I once lay in the grass having just come out of the lake and was muddy and smelled of the mossy green water when I heard a strange long plaintive call from the trees. The trees were filled with crows. The voices of two or more would stretch and bend, then blend together and be answered by crows all around, as in a kind of beautiful mourning.

One day as David and I walked in the woods a flat red sun burned through the trees. We came to the edge of the forest where it opened onto a broken old apple orchard we had never seen before.

The smell of damp earth, the sweetness of maples and pines in the air rising in the cool of evening impressed themselves upon me with a vividness that I remember with utter transparency, as though it were freedom itself.

There are three jewels in its forehead like kernels of amber,
and a white sticky powder on it at first. The joints of its wings
are turquoise. Its wings are wet and folded back like the ridges
of oyster shells. There are little fine hairs on its face and back.
Its feet are struggling hard now. The thorax heaves and curls.
The new legs unbend and feebly scratch.

     The cicada and the chestnut come out of their shells in the
same way. She makes a platform on the breakfast table for the
cicada to pull itself out of its shell. It bursts its shell very
slowly, by breathing. Like labor in reverse.

Out beekeeping today with Bob Kime. He tells me in winter
they dive into the sun-streaked snow, mistaking it for the sky
and skid across it on their wings until they die of
exhaustion or bury themselves, the little golden
bulbs of their bodies throbbing as they sting the air.

Today we swept them into clouds with our new air-gun
but they were too lethargic to react and mob us, and
they crawled into knotted heaps on the ground,
spinning up piles of dead leaves. We sat in our netted
helmets under the young beech trees and met some
hunters going by with terrible-looking arrows.

"I love making you talk.
Why, you could be run over by a hippopotamus
and you'd never say nothing about it,
n'less somebody said, 'Was you run over by a hippopotamus?'
and you'd say, 'Yeah. I was.'"

White-faced hornet nest in R.
Norway Maple.   8/11/93
Delancey Plaire

The white, thready form of an infant squirrel
was being eaten by the stream.
Day after day it trembled back and forth
over the bed of rock in the cold, wrinkling water,
that, drying up in the late spring,
never quite carried it away.

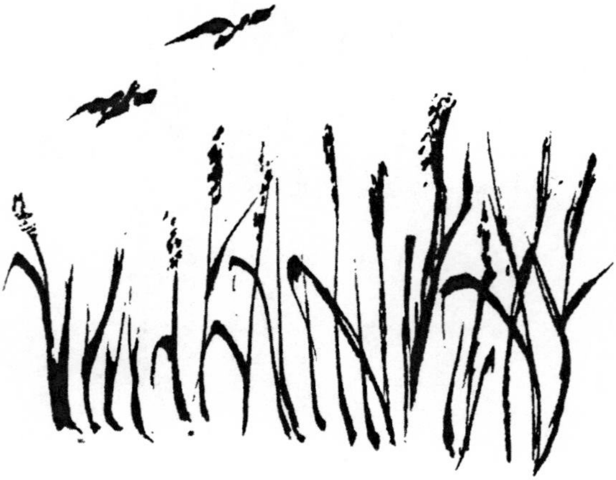

I saw an otter in the snow today
and crows on the ice
and thousands of geese.

Everything is blue and gold
and blue and gold
with berries stuck like drops of blood
to the trees.

Oh, Adam, shall I show you the tree of eternity
and a kingdom that does not pass away?

يا ادم هل ادلك على شجرة الخلد وملك لا يبلى

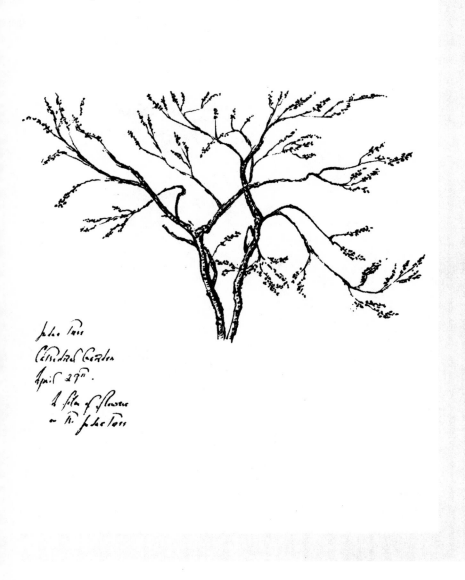

John Russ
Cathedral Garden
April 29th.
A sketch of flowers
— Rh John Russ

Hydnum
(edible fungi)

Clavaria
(coral fungi)

amanite
muscaria
(poisonous)

Cantharellus
cibarius
Chanterelle
(edible)

CLITECYBE
GIBBA

Polyporus
...

Tremella
... lutea
Very yellow

Boletus ed...

Auricularia
auricula (edible)

Spring Azure
yellow
orange
Buckeye
grey
orange
Hairstreak
Fritillary
little
wood
satyr
angle wing
nymphs
purples
mourning cloak

A cluster of trees with a crown of flowers
The birches by the pond in April

"Glorious isn't it—the trees. I love trees.
I could drive all day and look at nothing but trees.
Those sumacs a few days ago were bright scarlet.
Now they've had it.
'The sumac army climbed the hill . . .'"

Aunt Dorothy pulls my hand and with her cane goes
down the path to Otter Lake after some gorgeous
pink-red maple leaves.

"This is the last I'll see of the autumn leaves,"
she says. "Look at that desolate countryside—
all those rocks."

"I love to see the leaves on the ground. They look like
they're hugging the ground, keeping it warm
for the winter."

All of one's life is a beautiful pattern, like any kind of
growth. I have the image in my mind of a plant that eats
light and things invisible to us and, materializing them,
branches forever away from its core, yet is bound to it, and
has its ultimate predictable boundaries: invisible, yet where
the growth will go. Always in spasms it blossoms and
leaves in fragile fast-fading forms, then withers back into
itself, with only a tracery, the subliminal skin under the
new bark, of what was.

And as growth patterns, or anything formed in the physical
world gradually turn or collapse into stone, the earth's
skeleton, the energy that made them is freed, traveling
rapidly through and away as in a huge explosion, in which
we are some small part

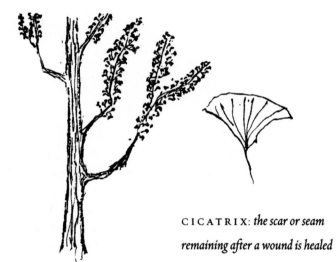

CICATRIX: *the scar or seam*
remaining after a wound is healed

the scar left by the fall of a leaf

The eye is a cicatrix
or umbilicus remaining
after the separation of
the umbilical cord from
the pericarp

4

GARY

"THE ONLY THING that's ever gonna hurt you, sweetheart, is people," Gary Lynch used to say. "An animal sees ya, they run."

In back of his house on the lake down our cottage row, Gary kept four or five foxes in plywood cages. I remember the shock of seeing them there, my first close view of foxes — their fur dimmed behind the crushed chicken wire, their strangely intelligent faces and lithe bodies gathered back like gnarled, badly knotted yarn in rage and fear, the quality of wildness almost fermented in them. Gary kept them in order to collect their urine for the lure he used to bait his traps. There was always some undercurrent wisp of animal smell on his clothes and body; I noticed it even the day that we first met, fishing for pike in the rain when I was sixteen.

It was an Elephant Lake day, as we used to say of gray, mist-laden mornings, like the mornings in Canada on Elephant Lake when we went out to troll in the weeds, and sometimes my grandparents came, my large authoritarian grandfather bringing white paper bags of walnut fudge from Haliburton, my grandmother reading to us from a French children's book as we sat for long hours in the rocking boat.

In the rocking boat on the empty lake in the mist my brother David, with a concentration he never broke to speak, stood casting in the bow over the rain-pocked surface of the water. I was in the stern, drenched, in my father's loose gray fishing trousers cinched with David's scout belt. David's new fishing companion was midboat, laughing in his soft Irish voice, his round face ruddy and smudged, his flannel shirt worn through at the elbows. He had an atmosphere of toughness and solid competence all about him, even on the stormy lake. In that soaking chill there was the sudden miracle of pulling up out of the water the slippery green pike, a tense shard of pure muscle gasping on the metal floor.

Northern pike were large unwieldy fish, with tiny sharp needle-like teeth. They were hard to hold in your hands and we often didn't bother to bring them in, for we rarely ate the fish. The flesh was too viscous and slippery and never cooked well, unlike the flesh of trout. But trout went out into the cool deep water in summer and could only be caught by the dull method of thermocline — letting out a hundred or more feet of heavy line tacked with silver spoons like thorned minnows. You couldn't feel the fish take the bait. There was only the laborious process of reeling in the line once in a while to see if there was anything on it.

The trout often came up with silver-blue sea lampreys stuck to their sides. All around the lamprey's mouth, a sharp, perfect circle of razors, the dappled skin of the trout was pale and flaking away. The lampreys came into the Great Lakes in the water holds of cargo ships nearly a hundred years ago and destroyed the trout fishing industry. I was thrilled when I saw what was then a rare migrant, the famous cormorant — the sea crow, an antique symbol of greed — for the first time, right off our dock. When a cormorant colony was smashed to pieces in the night on an island in Alexandria Bay north of us on the St. Lawrence in 1992 there was a national outcry — why would anyone destroy a colony of nesting birds? But we guessed what the cormorant had done to the bass fishermen who made their living there.

I sometimes slept on the dock and woke before dawn when David was already out on the lake. He would come in with bass we

would clean on the shore and cook for breakfast with butter and wal-
nuts and lemon grass.

When David died, Gary brought us fish. He brought hand-
churned butter and venison and bear meat from Pennsylvania, and
the black-tipped tail of a red fox. He said that one day he would bring
me a little ermine and a baby crow. "Do not despair. Angels are nearer
than you think." I sometimes thought of Gary.

It was not unusual to see a rabid bat fluttering around the house
in the middle of the day, like a scrap of black cloth in the wind. We
were taught to be afraid of rabies as young children, for it was the line
between the domestic and the wild, the thing that brought wild animals close. Once, out in Johnny Curry's green rowboat, I watched a
bat, oddly mustard yellow with fine leathery wings like large webbed
hands, flopping on the thin waves. Its soaked furry head peered up
humanlike as Johnny beat it under with the flat blond blade of an oar.
But for Gary dealing with rabies was routine. He would come over
and with one shot shoot the mothlike little bat out of the sky.

When I ran up from the dock where nesting swallows dove at
my head, Gary, sitting on the porch with my father, would laugh,
"Just go down there with a tennis racquet, sweetheart, and they won't
bother you."

Our fishing expeditions became a running joke.

"Rainwinds through the night. In the morning there is a stiff
wind and clouds. The light is still strong on the water as I take my first
swim."

"What are we liable to hook?"

"Oh — browns, rainbows, landlocked salmon."

"Landlocked salmon? What do they look like?"

"Torpedoes. Watch out for sharks."

"Are there landlocked sharks?"

"That was a joke."

"No, I mean, are there landlocked sharks?"

"Oh. Not here. Lake Nicaragua."

"Do you know that for sure?"

"Robin knows all about sharks . . . Run right the way you're running."

I am at the wheel.

"Jesus Christ! Did you see that! A big rainbow just come out of the water — see that swirl over there?"

There are vertical smudges on the graph. Gary says they are suspended fish. "Rainbows."

"Why do they hang that way?"

"Temperature."

Waves knock the boat. The boat rattles.

"I gotta hoe these beer cans out of the boat someday," Gary says.

It begins to rain. We are drinking black, bitter coffee.

"Yeah, I'm about the first one to use a downrigger on this lake," Gary says. "Everybody's still pulling copper. That's hell. If you're gonna fish it might as well be fun, right? The trick is, see, you keep the ball right off the bottom."

He holds up a round iron weight, painted orange. A knob on top forms the loop for the fishline.

"The ball pulls the lines, all the lines, down. Fish takes the line. The clip releases — clothespin effect."

"Wait a minute. If a fish takes the line, then the line goes free . . . ?"

"Free from the downrigger line. Then you haul 'em in."

There are four fishing poles stuck in plastic tube stands in the stern. Gary moves to the back of the boat to adjust the lines as we go shallow or deep, winding them up or down on spools attached to either side.

"If I spit in the water I can see how fast we're going," he says.

There are flecks on the graph right under the fuzzy line of the surface — trout. The bottom is almost smooth at ninety feet, with an occasional pucker down or up. A strong wind abruptly knocks the boat off course, to the west, and, as abruptly, the bottom line on the graph drops to a hundred twenty feet.

As we pass other boats the radio crackles.

"Lynchy, how deep you running your rig?"

"Ninety feet — running right on the bottom."

The radio again: "Keep it at ninety feet, right on the bottom. Lynchy got four on the downrigger."

Gary laughs, his face red. "See that! Every time you go to the back, they think you got a fish on."

He begins reeling up the line. I watch it on the graph. There begins a sudden drop that goes off the graph altogether, a deep drop in the floor of the lake. Then the line begins to rise again and is the outline of a hill ending in a small shelf. Near the edge of the shelf are small marks where the fish are.

"Hey come here! Hurry!" Gary calls.

He hands me a rod from one of the plastic tubes. The rod tip dances and wobbles and I begin to reel in what I can feel is a small trout.

"Laker. Dead weight," says Gary. "A rainbow'd give you a good fight."

The fish rises to the surface, then languidly spits out the spoon and flops away.

"Oh no!"

"The ones that hurt, Susan, are the ones that are ten pounds that you lose."

The rain has stopped. The water is black and the sun on the water has a metal shine. A bright crowd of gulls is floating by. A dense black mark appears on the sonar just beneath them.

"Baitfish, see?"

"Yeah?"

"That's why the gulls are there. There's trout underneath 'em. There's a dead one."

A bloated fish floats beside us.

"Someone drug that one up. What happens is — fish get the bends, see. Sure — if you drag 'em up from four hundred feet. You heard of divers getting the bends? Same with fish, you drag a fish up a hundred feet, it's bloated way out. That's why you take 'em up real slow — otherwise, well, what I do, I take a knife and stick it in the bladder to let some of the air out."

"Why are there trout right here near the point? Is it warmer, colder?"

"Protection. They can hide, and when the baitfish go by, they grab 'em. There are bigger fish in the lake that eat trout."

"Like what?"

"Bigger trout. Northerns'll eat 'em in the spring sometimes. But in the summer the northerns stay in shallow when the trout are out a hundred feet."

"Do northerns attack people?"

Gary laughs.

Gary was in love with the wife of a trucker who had long red hair and green eyes and was putting herself through medical school. I used to see her stringing up laundry along the side of their trailer in the afternoon, beside her garden of foxglove and asparagus and swollen leaves of lettuce that held the light. Across the road was the shack where old Joe lived, a shack made from doors, with a mean shepherd mongrel chained beside it to a tree. They must've put it down when he died. Now the place is full of grass and locust suckers, waist high, and musk roses bloom there. In the sumac nearby the foxes live. We hear them bark at night. In the field beyond, the winter wheat is white with gulls.

I saw a large stag in the hidden back field at the edge of the forest here. As I watched it, it watched me. It tilted its head back and tore the leaves from an old oak, its antlers like blunt broken branches. What will be points are still soft knobs covered with silver fur. Gary says the does can bite them now. This is why stags go around in groups in the summer. "The does can get back at 'em for pushing 'em around all fall." When the antlers get hard they breed. The stags keep their antlers until the last doe is bred. Then the antlers fall like ripe apples and the mice eat them.

Bob Kime said the stag I saw must be the "legend buck." Said he shot it with his bow three different times, and every time it walked away. "But wouldn't you know it, my wife hit it with the car — did fifteen hundred dollars worth of damage," before it dragged itself off the road and was gone.

When Gary came down to our woods in Hector, he would draw me in deeper among the trees, which were almost pavilionlike in the loose glacial till, forming a broad canopy above a stream banked with Solomon's-seal. He would stop and sniff the air and say, see there are minks denning here, in the lip of a gravel rise (their musk slightly sweeter than a skunk's but similar), or show me a fawn licked neatly apart and the track nearby, almost crumbled into the mud, of the animal that brought it down. Gary had a great love for foxes and the otherworldly redness of their fur. Sometimes in the spring we would walk quietly around a clump of sumac to hear the faint high yapping of their pups.

I didn't think much then about how a trapper had to notice things — the almost invisible roads pressed into the carpet of fallen leaves, the scratches on trees, the turned-up earth. I did sense that the knowledge Gary had — something between informed observation and instinct — was a rare and precious thing, a remnant of an older world, and that I wanted to have it too.

One day he brought a sable turkey feather he found standing upright in the field beyond our yard. It seemed incredible in the late 1970s that wild turkeys could be coming back. At the time, Gary had begun to see traces of a new animal raiding his traps, and he wanted to find out what it was. He took me out in the fields with a strobe light hooked to the roof of his pickup to see what was out at night. The colors reflected back from the mirror-like eyes in the dark told us at once what animals were there — deer green, raccoons orange, foxes red — before the outline of their forms emerged in the halo of light. But we never saw the animal he was looking for, though we watched and waited. "A coyote is an opportunist, see," Gary would say, as though to say he knew what it was to be an opportunist, to live on anything, to be subtle enough to utterly disappear.

We heard the farmers say, "I thought I saw a German shepherd in the field last night, but then you could see by the way it moved that it was a wild animal."

One Christmas morning Gary pulled up at my parents' house with six coyote skins, stapled through the eyes, in the back of his pickup. He brought them as a gift, proof of the animal, and how near it was. He had trapped five of them behind our cottage on the lake, and the sixth one on the Nielsen Road, a hidden field road where I loved to walk, and killed it with a stone.

We drove to Rochester one night, Gary giddy with laughter and exhaustion, his hands on the steering wheel crusted over with dirt. His eyes were bright behind his gold-rimmed glasses, for he had figured out how to trap this animal that had eluded him for so long. We went to see a friend of his, a furrier named Tom Monroe, who sewed the skins together into a coat. Tom laid the coyote skins out on his metal table, beside a newly acquired pelt of timber wolf from northern Canada, to show us how alike they were, the stiff black guard hairs where the shoulders would have been, the blond underfur streaked with white.

Gary used to say that the eastern coyote was really the Algonquin red wolf. He would have been gratified to see an article that ran in *Adirondack Life* not long ago, which flatly stated, "What we've been calling [eastern] coyotes here look an awful lot like the Algonquin red wolf."

The article went on to say that the wolf of Algonquin Park, long thought to be a smaller version of the gray wolf of North America, was really *Canis rufus,* the red wolf, placed on the endangered species list in the United States in 1965. The author noted a conclusion that many researchers have reached: that the red wolf is a hybrid that results from the interbreeding of coyotes and gray wolves. Perhaps this is the animal, the article suggested, that was described as a wolf in the Adirondacks in the nineteenth century, when "if the beast howled like a wolf and hunted livestock like a wolf, it was a wolf; taxonomy didn't much matter when the time came to pay a bounty."

In George Caleb Bingham's painting *Trappers on the Missouri,* a French trapper and his Indian son, their bag of shot ducks at their feet, drift on a misted river. A strange animal is tied in the bow. For years I thought it was a black cat, but one day I looked closely at the picture at the Metropolitan Museum, and saw that it was *Canis niger,* the black phase of the red wolf.

What is a wolf? Those that Audubon frequently encountered in his travels across North America varied greatly in size and shape and color. The term may be as elusive as the creature. In Greek the name of the animal belongs to the realm of words denoting light, like *lukos* (a name that comes directly from the word for light itself) and *argos,* silver (literally, "shining," used by Homer to describe the movement of canids, "because all swift motion causes a kind of glancing or flickering light," as the commentators Liddell and Scott have written), an image that makes me think of the dioramic wolf display in the American Museum of Natural History, in which wolves run across the snow at night, their backs the silver of the ground and the moon and the frozen lake.

In his 1992 study of the eastern coyote the Canadian biologist Gerry Parker points out that the small-bodied coyote of the American plains is the canid archetype, the compact blueprint that has morphed repeatedly over time into larger, more specialized forms. Parker tells of how plains coyotes followed prospectors north to Alaska a hundred years ago, drawn by the carrion of their dead horses and the garbage they left along the way. These northern-ranging coyotes bred with gray wolves on the edge of the boreal forest as the trees were cut back and the wolf population was dislodged. The new hybrid learned to hunt in packs in the hard winters, to bring down snowbound deer. As they changed from scavengers to predators they grew larger on a diet of fresh meat, and ultimately became the animal we see today. "What it is and how it got here," Parker concludes, "are the same question."

🐾

One snowless day last winter I saw a shred of white on the bare ground outside my window. I thought it was a piece of Styrofoam blown there by the wind. But later I saw the cat shaking it by its throat, and saw that it was a little ermine betrayed by what we used to call an "open winter," when nothing froze. The ermine turned white, but there was no snow.

The ermine, *Armenia mustela,* the Armenian rat, is the local weasel that eats the voles and mice around our house. I used to see the southern version in the streets of Cairo and in the open-air teashops on the banks of the Nile, where they would come up from their dens in the riverbank and stand on their hind legs beside the tables to beg for cake. "Ursa," people called it, saying "it eats blood." One time I went to see a rare bookseller in a denser, poorer part of the city. When the bookseller rolled up the metal door of his shop a little weasel bounded out.

The winter fur of the northern weasel was once called the purest white in nature, save for the tip of its tail, which is jet black as though dipped in ink. For centuries it was the badge of European royalty. Seeing it exposed and stranded in my own backyard made me think of Dick Wood's pocket trapper's manual from the twenties, which began, "Today wilderness conditions in this country and the Indians have about disappeared, but the fur-bearers remain."

The trapper and the trapped now meet as castoffs of the civilized world. The story of the devolution of the wolf into hostile predator, then into the scruffy eastern coyote, is not unlike that of the American trapper, the primordial hunter of the wild, who allied himself with the Indians, and so learned to survive without the hampering luxuries of settled urban life. The American trapper is something of a Caliban figure now — a chthonic, scruffy character, outmoded and despised. But Caliban is, after all, the native of the place, the one who knows where everything is. As he says:

> I prithee let me bring thee where crabs grow
> and I with my long nails will dig thee pignuts,
> Show thee a jay's nest, and instruct thee how
> to snare the nimble marmoset. I'll bring thee
> to clustering filberts, and sometimes I'll get thee
> young scamels from the rock. Wilt thou go with me?
>
> (Shakespeare, *The Tempest*)

A man walks up to a woman in line beside me in Penn Station and demands to know, "Is that real fur?" "I'm afraid so." She blushes, catching her breath. He smiles triumphantly. I look down at his shoes. They are made of leather, a material so pervasive in American society that one forgets it is skin, a word that means to cut, to flay, an animal that has not even been caught in the wild, but raised to be killed. That leather is skin is unimaginable, forgotten, but fur still seems to be the animal itself.

That was the idea. The American Indians were matrilineal and totemic: they derived their identity not from who their father was, but from the animal totem of their mother. Totem is an Algonquin word meaning kin, brother or sister of the same womb, suggesting at once the palpable sense that the womb, the mother, was the origin of life, and that all living things had this origin in common. The difference lay only in the manifestation, the form.

The belief in the common nature of all life arose from the close study of life itself, from the informed eye of the primitive hunter. To capture an animal — to outwit it with a trap of sticks and stones, or come close enough to catch it with one's hands — one has to know it, and to know an animal is to know that one is an animal oneself.

"As for scalping or even skinning a savage," says the trapper Hurry Harry to the Deerslayer, "I look upon it pretty much the same as cutting off the ears of wolves."

The line reveals the odd correspondences of the early North American fur trade: that the European saw the Indian and the animal as one, belonging to the wild; that the European took the practices of the wild — scalping and trapping — from the Indian; and that scalping and taking the fur of animals were similar practices, with similar intent. It was the loss of the intent — the wearing of the vestiges of animals and enemies as a badge, something invested with significance, a form of affiliation — that created the environmental disaster of the commercial fur trade, the impetus for the European settlement of the entire North American continent.

🐾

The closeness of human and animal life was part of the mystery of North America. In *Apologies to the Iroquois* Edmund Wilson conveys it even in the late 1950s, when he tells of walking into a modern kitchen on the Tonawanda reservation in western New York, where Seneca men have gathered to perform a sympathetic healing ceremony called the "little water." The little water is the miraculous ointment that brings the dead back to life. A common element in fairy tales, such an ointment is so precious and rare, so hard to obtain, that it is kept in a tiny hidden vial of crystal or diamond. In the Iroquois version the diamond vial is an acorn, and the ointment is made by the animals of the forest from their own hair, mucus, and blood.

Wilson describes the small group of Senecas wearing western clothes casually assembled in an ordinary American scene of the evening kitchen. A television drones in the next room. The lights go out and the chanting begins, with the men, indistinguishable in the dark, taking the parts of the different animals.

As the animals find a hunter in the forest scalped and dead, a wolf comes first to lick the blood from the exposed flesh of the hunter's naked head. The animal voices — the birds' cries, the throaty growls of wolves — were so accurately produced in the room, it was as though, Wilson says, in a description that makes the hair stand up, the actual animals were present in the dark kitchen, and their presence was the medicine itself.

The territory that became New York State was among other things a scene of historical compression and wild instances of mistaken identity, where Stone Age and Iron Age collided like tectonic plates. When John Cabot sailed to the Canadian Maritimes in 1497 he was looking for the coast of China. Cabot had been to Mecca in the spice trade and had come to believe that a way to get around the Arab traders, who acted as middlemen on the Silk Road and took extravagant cuts, could be found in the direction farthest away from them — north and west. When Champlain traveled the St. Lawrence to Lake Ontario by canoe in the early seventeenth century, he still thought he was in the outer reaches of Asia. Jean Nicollet, one of Champlain's forest scouts, set off across Lake Michigan with a gift for the emperor, a robe of damask, thickly embroidered with birds and flowers in gold thread, rolled up in the bow of his birchbark canoe. Sixty years later La Salle's men, weary and disgruntled, turned back on the same route and settled along the St. Lawrence near Montreal in a place they cynically called "la Chine."

What the early explorers hoped to find in China was more direct access to the luxuries of the East, for exotic luxuries were the only things their sponsors, the kings of Europe, were interested in. It took a hundred years for the European aristocracy to accept the rich fur from the depths of the uncut American forests as a luxury that rivaled silk.

Gary Lynch, like Bob Kime, belonged less to the history of the mass industrialization of the American continent than to the prior topography of New York State, its wetlands and forests. They worked at a lost connection as intermediaries between the settled and the wild, tracing a vertical line, backward in time:

A desperately fluttering bird, tied by its feet to the lower branches of a pine, might draw a lynx into a fiber snare tucked into the dark shadows beneath the tree. A deadfall — a rock propped over soft meat — makes a natural cave for a fox. The unusual brightness of split green sticks, stuck in the ground to form a pen, will occasionally lure in a curious black fisher.

Like the wolf, the beaver was one of the primary totems of the tribes of the Northeast, of the nomadic Algonquin, and of the sedentary agriculturalists of the eastern Great Lakes, the Huron and the Iroquois. Members of a clan claimed their descent from a giant version of their totem animal, like the grizzly-sized beaver that inhabited the early Cenozoic marshes of these parts; the totem animal was normally a familiar one, routinely hunted and killed for its flesh and skin in an intimacy of exploitation and identification and absorption.

Killing a totem animal required rituals of propitiation. A passage in *The Golden Bough* gives a sense of the elaborate precautions involved:

The Canadian Indians were equally particular not to let their dogs gnaw the bones, or at least certain of the bones, of beavers. They took the greatest pains to collect and preserve these bones, and when the beaver had been caught in a net they threw them into the river. To a Jesuit who argued that the beavers could not possibly know what became of their bones, the Indian replied, "You know nothing about catching beavers and yet you will be prating about it. Before the beaver is stone dead, his soul takes a turn in the hut of the man who is killing him and makes a careful note of what is done with his bones. If the bones are given to the dogs, the other beavers would get word of it and would not let themselves be caught. Whereas, if their bones are thrown into the fire or a river, they are quite satisfied, and it is particularly gratifying to the net which caught them." Before hunting beaver they offered a solemn prayer to the Great Beaver, and presented him with tobacco; and when the chase was over, an orator pronounced a funeral oration over the dead beavers. He praised their spirit and wisdom. "You will hear no more," said he, "the voice of the chieftains who commanded you and whom you chose from all the warrior beavers to give you laws. Your language, which the medicine-men understand perfectly, will be heard no more at the bottom of the lake. You will fight no more with the otters, your cruel foes. No, beavers! But your skins will serve to buy arms; we will carry your smoked hams to our children; we will keep the dogs from eating your bones, which are so hard.

In Europe the beaver was no longer an animal. It was a hat. Beaver fur became felt, sold not by the pelt but by the pound. The hatters of Paris and London found beaver fur to be a felting material superior to wool. The hairs had a toughness that made them at once naturally cohesive and somewhat waterproof. America provided what seemed to be an inexhaustible source of felt. A hatter would painstakingly pull out each guard hair with tweezers, then brush a solution of nitrate of mercury over the plucked fur. The mercury scaled, or roughened, the shafts of hair, helping them to interlock in the felting process. The mercury also poisoned and killed the hatter. "Mad as a hatter" referred to someone who, after long exposure to mercury poisoning, had lost muscle control and the ability to speak. The hatter played a violin over the shaved hairs to create a steady vibration that would gently shake away the dust before the hairs were matted into a dense felt, dyed black, and shaped. The fur of the beaver became the top hat, an essential fashion accessory in Europe and America for over two hundred years. One can hardly imagine the millions of animals killed simply to sustain this fashion.

To harvest the beaver meadows that lined the water routes inland from the Maritimes, it was necessary to travel through densely forested country, and that could be accomplished only on water, by means of tree bark canoe.

Champlain was the first European to master this instrument. Thus an elegantly eccentric polymath began the tradition of the American trapper. Champlain, precursor of the Enlightenment in the American wilds, studied his surroundings — animals, plants, weather — even when he was stranded in the Haliburton Highlands or menaced by the hostile Seneca on Oneida Lake. The Huron and Algonquin loved this quirky alien who studied their unwritten dialects, kept elaborately illustrated journals, grew Mediterranean roses in Canadian permafrost. He companionably set out with them to fight the Iroquois on the shores of Lake Ontario. Fantastic as the Tin Man with his suit of French armor glinting in the sun, Champlain's very appearance routed the enemy immediately. "The mosquitoes," he wrote afterward, "were terrific."

Although Champlain's interest was primarily in exploration, his funding came from the fur trade, which compelled him to seek the best tribal sources for fur. He sent young Frenchmen to live with the tribes — not to teach, as the Jesuits did, but to learn the tribal languages, knowledge of animals, methods, routes of travel. These young men, often teenage boys who disappeared for years at a time, were the *coureurs de bois,* the forest scouts, or, as the French dictionary translates the term today, "trappers."

The Jesuits ultimately failed to win over the tribes and their land, though they secured for themselves the martyrdom of saints, an ordeal of sanctity reverently set down even as the priests' fingers were severed, one by one, with the shiny pink edges of freshwater clamshells. The *coureurs de bois* succeeded, silently, in opening up the entire American continent.

In the seventeenth and eighteenth centuries thousands of young men, many of them European aristocrats, vanished into the American wilderness to make their way as solitary trappers. The tradition and its attendant mythology became so entrenched that when Audubon, painting in the artistic conventions of Central Asia, was unable to sell his work in America, he went to Europe and presented himself as an American woodsman in a wolfskin coat.

At the time, the figure of the American backwoods trapper from upstate New York, Natty Bumppo, was one of the most popular fictional characters in the world. But by then it was becoming unfashionable to deal in furs. Fenimore Cooper made Bumppo a rifleman. The Susquehanna trapper Hutter is the bad guy in *The Deerslayer,* scalped like one of his otters or beavers in retribution in the end.

For trapping went out of fashion with the beaver hat when the beaver meadows were exhausted in the nineteenth century. Once numbering in the hundreds of millions across North America, the animal had virtually vanished from the East. In 1907 Theodore Roosevelt's Department of the Interior released fourteen beavers from Yellowstone in the Adirondacks in an attempt to reestablish the beaver in New York State.

❧

When Gary lost his trapper's license for setting a traditional rope snare instead of using the plastic leg-hold trap the law requires, he spent the summer growing roses, like Champlain.

A summer or two later Gary died. He drowned while working for the telephone company, swimming a fiber-optic cable across the Erie Canal in Lyons. Normally his work crew would have hauled the cable across in a rowboat, or would even have tied a light rope to an arrow and shot the arrow across the canal, then used the rope to pull the cable along. Both methods seemed too complicated and time-consuming to Gary. The weather was hot, good for a swim. He stripped to his underwear, tied the rope around his middle, swam two thirds of the way across — and then, inexplicably, went under. He came up to the surface, shouted, "Hurry! Hurry!" to his men, and went under again.

Other linemen dove into the canal but could not find him. The fire company came an hour later and hauled his body out with a grappling hook. My mother called the Hudson Valley farm where my husband and I now live, and Lan and I drove the long familiar road back to Geneva.

I walked into the field across from Chuck Brust's and found a red hawk's feather standing in the wet grass beneath the hickory tree where I once saw a redheaded woodpecker and made a note of it in a pocket notebook for the first time. I used to come here in the evening to watch the deer.

A week before he died, Gary told me on the phone that a fox smells like a skunk, and a coyote sweeter than a fox, and that he would come and stay with us on the way out to a job for the phone company on Long Island. Our freezer was filled with the lake trout he had caught that summer.

We drove out to Lyons. The canal was narrow there, and we could see how Gary thought it would be easy to swim across. The lockmaster, a leathery man with tattooed arms, said he thought Gary could not simply have drowned, but must have had a heart attack.

"We're trying to make sense out of it," we said.

"You'll never make sense out of it," he said. "It's a tragedy."

In the morning, before Gary's funeral in Seneca Falls, I went down to the dock on Seneca lake, where blue was breaking through the fog in an iridescent haze. There were whitecaps to the south in a silver edge of spray. Wind out of the south-southwest. At the house on Delancey Drive in town, as we prepared to drive over to the funeral, we saw a white-faced hornets' nest in the Norway maple in the front yard.

"It looks like a heart," Lan said.

Bob Kime promised he would come that night, when the temperature had dropped and the hornets were asleep. He would take down the nest, hornets and all, put it in a plastic bag in the deep freeze, and sell it to a research lab in Washington State. It was good to hear his voice.

5
AN ATMOSPHERE
OF SWEETNESS

ANYONE LOOKING at the Finger Lakes region and its strangely dense history must wonder what explains so many powerful phenomena arising in what would otherwise have seemed a backwater. What odd metaphysics brought together Mormonism, spiritualism, women's rights, abolitionism, the origins of the American fortune in the fur trade, and the scientific advancement of agriculture?

I once copied out the development of the word "talisman" on three index cards taped together and stuck them on the wall above my desk. They read:

> **Talisman:** an object marked with magical signs and believed to confer on its bearer supernatural powers or protection. 2. Anything having apparently magical powers. French and Spanish *talisman* from Arabic *tilsam* from late Greek *telesma*, completion, consecrated object, from *telein*, to fulfill, consecrate, from *telos*, fulfillment, result.

What would be the talisman that one could follow through the layers of Finger Lake soil, through layers of memory and history, from the Iroquois to the Experiment Station? What would signify the indelible native atmosphere? It might be the apple, or its essence, sweetness itself.

In the basement of the Warren Smith Library at William Smith College across the street from our house in Geneva not long ago, I asked the archivist for a file of William Smith's automatic writing. Warren Smith's great-uncle William had been an avid spiritualist, and initially tried to found William Smith as a spiritualist college. But the open-ended, borderless ambitions of spiritualism ran counter to formal education, or any hierarchical structure at all. William Smith couldn't figure out how to put the two together.

The archivist had lived around Hydesville. She said the fire department there decided to burn down the Fox sisters' house ten years ago as an exercise to keep the firefighters in training. She offered to drive me out to the weed-choked field where the house once stood. We might find a trace of the books and papers that must have been inside.

William Smith came to Geneva from rural England in 1840 to work as a farm laborer. He and his brothers bought land around the ruins of Kanadesaga and became pioneers in the American nursery business. They discovered that the rich clay loam, left like sludge around the meltwater of the glacier that formed Seneca Lake, was particularly good for growing apple trees and a plant that is botanically a simpler manifestation of the apple, the rose.

By 1870 the Smiths and other nurserymen had four thousand acres under cultivation, creating in the terraced land above the lake some of the largest commercial orchards in the country. The Smiths planted yearly twenty thousand rose bushes in over two hundred varieties on a single acre. In summer the smell of the rose fields filled the air of the town.

The Timber Culture Act made it mandatory for settlers moving west to plant trees in order to claim land: to plant an apple orchard meant to own a farm. In the decades after the Civil War, the Smiths sent daily shipments of grafted apple trees west across the country on the new railroads to meet the demand, shipping as many as six million pounds in a year.

William Smith traveled around the world in search of new apple seeds and stock, and surrounded his house on Castle Street with a twenty-five-acre botanical park of unusual trees and plants that he collected on his travels. The trees were not merely a business, but an enactment of Smith's interest in the natural world that had its roots in a mix of spiritualism and Swedenborgianism.

At the time, the writings of Emanuel Swedenborg saturated American thought in its transition from the abstraction of America as a promised land to the idea that the landscape itself was numinous, a transformation from Jonathan Edwards's man as a spider suspended over the fires of hell to an Emersonian luminosity: now the fire was within.

Swedenborg's early work involved mineralogy and other areas of hard science. He studied widely, and at Cambridge attended Edmund Halley's lectures in astronomy. Following the model of Halley (who had made his stunning discoveries by observing the heavens minutely), Swedenborg investigated the question of the exact placement of longitude, the physical nature of time. The work of closely examining the night sky occupied Swedenborg for forty years. His long immersion in objective scientific observation seems to have opened into a vision. He saw the New Jerusalem of Revelation come down from heaven and become manifest on earth as a geometric plan, a square made of luminous particles — like stars or precious stones — with doors on every side, a pattern imbued with divine light, only mankind could not see it. The doors were always open, and within there was no night, for everything shone with the soft jewel-like light of the presence of God. In the heart of the New Jerusalem stood the tree of life, bearing fruit every month of the year, and all who entered could freely partake of the fruit of the tree.

"Nectar and ambrosia," Henry David Thoreau wrote in the 1840s, "are only those fine flavors of every earthly fruit which our coarse palates fail to perceive, just as we occupy the heaven of the gods without knowing it." Heaven was earth, suffused with the illumination of divinity.

In the 1820s the Swedenborgian missionary John Chapman —

American folklore's Johnny Appleseed — wandered west through the Ohio River Valley. Chapman lived like an Indian, carrying little. He wore a shaggy beard and slept unsheltered in the wild. Everywhere he went he planted apple seeds so that within the unlimited expanse of the New Jerusalem all would eat of the tree of life, which in his European conception of the Garden of Eden was the apple tree.

The apple is not native to the ancient Near East with its tradition of the cultivated garden or "paradise." The native Mediterranean "apple," or archetypal fruit, is the fig, whose connection with birth and the feminine is self-evident and translates into the contemporary Italian obscenity *fico* with its obvious derivatives. The Greek word for fig was *sukon,* from which derive words that mean sweetness itself: sugar, suck, as well as sycophant, a displayer of figs, a pimp. In the biblical tableau of the Garden of Eden the woman bears the Semitic name Eve, *hawa,* the word for life. *Haya,* the snake, completes the visual pun. The snake in the tree of life is life itself, ever renewed, the current of life that flows through the female body, the flower and the fruit.

"The apple of America is a totally different apple," Henry James said. His father was a Swedenborgian spiritualist who had had a religious conversion in western New York, where the James family had amassed a fortune by feeding the workers who dug the Erie Canal.

In keeping with his interest in spiritualism and the natural world, William Smith built an observatory in his botanical park. He persuaded William Brooks, an Englishman famous for the discovery of comets, to direct its construction and become its resident astronomer. The observatory was a round, spiraling building, with a round library paneled in dark wood below, and beside it a small transit telescope for calculating time by the movements of stars in a narrow north-south patch of sky. The huge celestial telescope on the floor above weighed eight thousand pounds and sat on wheels that swung it around beneath a silver dome. Brooks moved the telescope clockwise by pulling a rope with considerable effort; with another rope he slid back a panel of the dome to open it to the sky. For a decade or more, Brooks lived for the most part in the observatory itself, where his constant vigilance enabled him to discover fourteen new comets and win the Leland Medal from the Academy of Science in Paris.

🌿

William Smith was a rugged-looking man with a shock of white hair and a thick white beard on his hatchetlike, almost Indian chief's face. He never married, but his closest friends in Geneva were women — the notorious radicals Elizabeth Smith Miller and her daughter Anne. Elizabeth Miller was the granddaughter of John Jacob Astor's partner in the fur trade, Peter Smith. She had grown up in her grandfather's house, east of Skaneatales, when her father, Gerrit Smith, was one of the wealthiest men in America.

The Smith manor house in the days of her childhood was filled with people from New York and Boston seeking her father's patronage, with the displaced remnants of the Iroquois tribes who knew it was an easy place to find food and money, with runaway slaves on the Underground Railroad, and religious eccentrics seeking refuge in western New York. Smith's niece, Elizabeth Cady, who fell in love there with one of Smith's protégés, the abolitionist Henry Stanton, wrote that her visits to the manor were the happiest days of her life.

Elizabeth Miller and her daughter eccentrically wore a Middle Eastern costume of trousers and a long shirt around Geneva. Amelia Bloomer, the editor of the reform paper in Seneca Falls, wrote that if women across the country would adopt this costume they would be freed from the gradual suffocation of the corset, a critical step in the freeing of women themselves.

In 1904 Elizabeth and Anne Miller persuaded William Smith that although he was unable to formalize spiritualism into a completely new kind of education, the most innovative thing he could do would be to found a college for women as the twin of Hobart, the conservative Episcopal men's college in the town. But William Smith College would represent intellectual freedom — it would have no connection with traditional religion.

Spiritualism, was, after all, a women's movement — something that took place anonymously, in the home, under the direction of women — by its very nature antiauthoritarian. It involved the haphazard, personal investigation into the essence of religion itself — the meaning of death — and though it was essentially a scam, it was also essentially true.

It can accurately be said that everything is made of light, and that death and decay are nothing more than the freeing of trapped energy. The continuous motion of particle waves that constitute both free energy and energy trapped in matter manifests as light.

The human eye perceives only a fraction of the vibratory spectrum that creates the closely linked entities of light and form, and sees a world of moving patterns as a solid thing. This perceptually defined conviction in the solidity of a world of moving patterns has long given rise to the sense that there is a hard barrier between form and formlessness, between life and death, a barrier that is everywhere feared as something dark. Yet there has always been a parallel intuition that death involves crossing into a realm of light, and that beings of light inhabit some kind of other world.

Images and ceremonies prompted by this intuition abound in every culture and religion. Some are of extraordinary beauty. The dissolution of consciousness witnessed by a tantric practitioner at death is perceived in stages as waves of color — the appearance like fireflies in this progression oddly suggesting the particle wave spin of the freed electron. Fireflies, like stars, have long signified the world beyond.

Heaven was originally precisely that: the starry sky, dating back to the earliest Egyptian texts, which include magic spells that enable the soul to be sewn in the body of the great mother, Nut, literally "night," like the seed of a plant, which is also a jewel and a star. The Greek Elysian Fields derive from the same celestial topography: the Egyptian "Field of Rushes," the eastern stars at dawn where the soul goes to be purified. That there is another, mirror world, a world of light, and that this world is simply the sky — and a step further, the breath of the sky, the weather, the very air — is a formative belief of great antiquity that has continued to the present day with the godhead becoming brightness itself: *dios/theos* (Greek); *deus/divine/diana* (Latin); *devas* (Sanskrit); *daha* (Arabic); *day* (English).

Madame Blavatsky unknowingly followed the pattern of Manicheanism, cobbling together a relevant contemporary religion from fragments of ancient and modern belief. Predominant among them was the unshakable belief in the spirit world, in the luminous souls of the dead.

Blavatsky's success was extraordinary. Theosophy rapidly spread around the world with disciples as diverse and powerful as Gladstone and William Butler Yeats, and opened the way, it has been said, for a new strain of Buddhism in America in the twentieth century.

Madame Blavatsky founded the Theosophical Society in New York in 1881. It was an amalgam of eastern esotericism and spiritualism; she had, until that date, been active as a spiritualist medium. The disarming thing about her is that she saw theosophy as a vast and delightful joke which she played at the world's expense, and if, incidentally, she was enabled to get rich . . . well, why not? "If personal sensibilities can be trusted, she is a genuine being with a real desire for the good of mankind," wrote Henry Sidgwick of her not long before she was unmasked . . . by the Society for Psychical Research, of which he was a founder.

(Ruth Brandon, *The Spiritualists*)

Blavatsky wrote of herself:

> What is one to do when, in order to rule men, you must de-
> ceive them, when, in order to catch them and make them pursue
> whatever it may be, it is necessary to promise and show them
> toys? Suppose my books and "The Theosophist" were a thousand
> times more interesting and serious, do you think that I would
> have anywhere to live and any degree of success unless behind all
> this there stood "phenomena"? I should have achieved absolutely
> nothing, and would long ago have been pegged out from hun-
> ger." (Ruth Brandon, *The Spiritualists*)

Phenomena tended to be simple tricks like bedsheets and bells
swung around on strings in the dark. Blavatsky claimed to have used
them to suggest the reality of an intangible world.

Pinning the spirit world down was a difficult task, as William
Smith found in attempting to create a program in which it could be
taught. Sherlock Holmes liked to say that he had figured out his own
eccentric education, a mix of chemistry and tracking, by himself. He
had invented his profession and was its only practitioner. Sir Arthur
Conan Doyle traveled widely in America and Europe to lecture on
the new religion, and he seems to have modeled the character of
Holmes in part on the spiritualist investigator: the person who was
called in to figure out the trapdoors, false-backed closets, and hidden
compartments in a dark room. Conan Doyle's best friend was a
rabbi's son, Harry Houdini, the spiritualist investigator par excel-
lence, who ultimately used the devices he uncovered as stage tricks.

When William Smith died in 1912 his nurseries were still supply-
ing the ornamental trees and shrubs for Central and Riverside Parks
in New York. But the nurseries soon fell into neglect and filled with
weeds. Over the decades the town gnawed away at the orchard land.
What was left of the nursery business went into partial bankruptcy
and was sold to a catalog company.

Smith left his house and observatory to Hobart, but the college considered the house, way up by the ruins of Kanadesaga where the nurseries were, too far from their location on the lake and sold it almost at once. The botanical park was broken up and sold off in housing lots. The observatory was too expensive to disassemble and cart away. For more than half a century it stood open to the rain and wind. The library, papers, and celestial photographs within were taken to the town dump.

The other day an old friend of mine who bought the observatory as a weather-beaten ruin thirty years ago showed me where he made a break in the water-rotted wall of the room that still contains the great celestial telescope. Behind the broken wood he found the rounded walls thick with honeycomb, honey, and bees.

🐝

What remained of the vast nursery land became the New York State Agricultural Experiment Station, where, during my childhood, Terry discovered the molecular structure of sweetness. He once wrote:

> In May of 1963 I met Robert Sands Shallenberger for the first time in a bar on University Avenue in Berkeley. He talked about the subtle and complex structure of sugars with such passion that I was thrilled when, after several drinks and many marked-up napkins, he asked me to be a student in his lab. Starting from Shallenberger's notion that the initial event in the perception of sweetness was the hydrogen bonding of a stimulant to receptors on the tongue, we developed a model we called "the A H-B Theory" of sweet taste, published in *Nature* in 1967.

The perception of sweetness was a process of absorption, like color vision, that depended as much on the molecular configuration of the membrane of the receptor as on the composition of what was to be perceived. The taste of something sweet involved a "vibratory hydrogen," a hydrogen atom that would jump to the electronegative center of the receptor, creating a weak hydrogen bond. Terry envisioned the process in the form of an electric plug, where the positive charge within the sugar loses a hydrogen atom to the negative charge within the molecular structure of the receiving membrane, and pulls a loose electron from the receptor into its own molecular sphere. The sweetest-tasting thing, like the fructose in an apple, was the chemical compound with the greatest solubility, the ability to flow and bond with the receptor tissue lining the tip of the tongue.

I never knew that this was what Terry actually did for a living. Although I remember when he left the sense of taste for the sense of smell around the time I moved to New York to go to school. Taste told you only one thing, he said, but smell told you everything. He would say things like, "You know that characteristic smell of a New York apartment? I figured it out in the lab today: cockroaches. You see one, there are a million in the wall." Or when I came home from New York for the first time and told him how I loved the smell of rain. "The smell of beets," he said, meaning wet ground. "I'm working on the patent as we speak."

Terry was figuring out in those days how to isolate the precise molecular components that constitute a particular smell. Then he invented a machine to help him track them down.

This was the kind of thing that people did at the Experiment Station. I thought of the Station as a perfectly ordinary institution. The director was a friend of my parents and a member of our church, as were Anne's parents, Ed and Nell Glass. They were the Station to us: Nell's bright blue eyes, her Southern voice, the full-blown fragrant pink roses on her table from summers long ago. She liked to say that she remembered how I would stand at the end of our cottage row waiting to see their blue Oldsmobile coming down the road on the way to our house for supper.

Dr. Glass was a tall man, with a spare New England manner and a voice that vibrated slightly down in his throat, having an odd tremolo quality. I remember going as a child with my mother to see him in the entomology department at the Experiment Station, where

he worked for almost half a century, a visit that gave me a lifelong
love of collecting, of glass-faced boxes and cases, for in the lobby
there I saw with delight for the first time in my life, mounted on hidden pins that pierced a kind of tucked white cotton, shimmering but-
terflies that seemed to be made of bright colored metal, staghorn beetles, and other strange oversized insects. We may have been bringing
him some childhood discovery from the cottage — a walking stick,
segmented like a twig, with antennae that curled back like wisps of
bark torn from a tree.

"But insects are so much like plants, aren't they?" I used to say to
Dr. Glass.

And he would laugh, "I can't see how," his pale blue eyes bright
behind his thick black-rimmed glasses. "How?"

"The way they reproduce."

"They both reproduce."

"I mean the *way* they reproduce," I say, thinking of how insect
wings uncurl and dry and leaves uncurl and dry, but I can't ex-
plain it.

Critical discoveries had been made at the Station for over a century. There were also notorious incidents of academic malice. A history of the Station published in the late 1990s tells of how, forty years
before, a young researcher named Robert Holley began to see something unusual in his work on the molecular composition of New York
State grapes. A supervisor forced him to drop the project. The researcher quit and pursued his work elsewhere. In 1968 Holley won the
Nobel Prize for the discovery of RNA.

When I was growing up, Dr. Glass was often away in India or Africa or South America solving insect problems. These were essentially
problems of food production. In response to Rachel Carson's *Silent
Spring,* Dr. Glass and his colleagues spent thirty years trying to figure
out how to control insect pests responsible for the wholesale destruction of crops around the world, without the use of poison. They
eventually figured out how to scramble pheromones, the chemical

signals of insects, so that the insects could not reproduce. The immense patience involved must have seemed like going backward.

The Station began as a place where chemicals were developed and monitored to improve food production. Its work had its origins in the birth of chemistry as an academic discipline a hundred and fifty years ago, in the "agricultural chemistry" of Justus von Liebig's *Organic Chemistry and Its Application to Agriculture and Physiology.*

🐝

Walt Whitman took the unusual phrase *Leaves of Grass* from Liebig's writings. Abraham Lincoln studied Liebig and envisioned a nation of small family farms, operated without slave labor by educated farmers who could add chemical fertilizer to their land to dramatically increase the yield of crops.

The first scientist appointed to lead Lincoln's newly created Department of Agriculture, in 1862, was Liebig's student Charles Wetherill. The New York State Agricultural Experiment Station was established in Geneva two decades later. At the corner of Castle Street above the Station land stood a farmhouse in the shape of an octagon, like a seal representing the geometric harmony of the molecular underlay of Liebig's agricultural chemistry.

🐝

"Agricultural chemistry" was redundant. *Chem* meant agricultural land. It was the hieroglyph for Egypt. *Chem* was the black land, as opposed to the barren red land of the desert, the rich arable soil left by the receding waters of the Nile flood, where agriculture and its necessary contingent mathematics — counting, the essential instrument of prediction — began.

The blackness was the raw soil itself, the wet black mud oozing out of the receding flood as the air all around rang with the cries of water birds — pelicans, herons, flamingos, the snowy egret on its luminous wings — wheeling down to feed on the myriad stranded living things. The sacred ibis was the most spectacular — a bird hunted out of the papyrus swamps a hundred years ago for purple hat plumes (though I saw one once, in Khartoum, along the river from under the mahogany trees, among the leaves like clusters of pale green grapes, a brilliant purple bird blotched with white slowly spinning down onto the mud flats of the White Nile to feed).

In dynastic Egypt hunting the sacred ibis was punishable with death. The bird was the god Thoth, who invented letters and numbers, and is always shown with a pen and a notepad in his hand. What the bird signified in that desert country was critical to know: when the water would come, and when it would go, and what the configuration of the stars would be at these pivotal times. To know these things one would have to see patterns, and create patterns — to measure, with absolute precision, distance, quantity, and time.

Mathematics, measurement, chemistry, emerged from the chaos of a wetland that mirrored the profound order of the stars. In this wetland a water lily was the notation for a thousand; a tadpole, a hundred thousand; and the frog, Hecate, the sound of the night itself — *hecate hecate hecate*, the mark meaning to repeat, to double the quantity again and again.

Thoth passed through the ages as Hermes Trismegistos ("thrice great," a standard hieroglyphic epithet: three water birds walking together in a line). Alchemy, "the Egyptian thing," arose from the attempt to recover the heretical knowledge of measurement and proportion that was lost in antiquity.

Pythagoras established a religion along the lines of the hieratic discipline of mathematics that he learned in Egypt: everything reduces to the eternal truth of mathematical formulae, an ethereal realm of pure form. His followers Democritus and Plato understood atomic theory in much the modern sense. Their notion that the universe is composed of an infinite number of radiant particles combining into patterns in a predictable flow of movement one might imagine arising from the infinite number of patterned stars flowing across the sky. The theory fell out of favor for two thousand years because Aristotle did not agree with it. The name coined for the indivisible, ultimate nature of matter was *a-tom*, "that which cannot be cut up," for everything else could be broken down into invisible particles that arranged themselves with geometrical harmony and precision into all the different forms of life.

The visionary recluse Lucretius urged his readers not to be blinded by religion (not to be "tied up" — the literal meaning of the Latin word *religio*) but to see with their own eyes the world that is before them and not to fear death, for all matter is composed of radiant indestructible particles called atoms that combine, come apart, and recombine. *The Nature of Things* begins with an invocation to the creative principle of the whole earth and the richness of the soil itself, the sweetness of sexual love:

> Alma Venus, caeli subter labentia signa
> Quae mare navigerum, quae terras frugiferentis
> Concelebras — per te quoniam genus amne animantum
> Concipitur visitque exortum lumina solis —
> Te, dea, te fugiunt venti, te nubile caeli
> Adventumque tuum, tibi suavis daedala tellus
> Summittit flores, tibi rident aequora ponti
> Placatumque nitet diffuso lumine caelum.
> am simul ac species patefactast verna diei
> Et reserata viget genitablilis aura Favoni,
> Aerieae primum volucris te, diva, tuumque
> Significant initum, perculsae corda tua vi;
> Inde ferae pecudes persultant pabula laeta
> Et rapidos tranant amnis (ita capta lepore
> Te sequitur cupide quo quamque inducere pergis);
> Deinque per maria ac montis fluviousque rapacis
> Frondiferasque domos avium camposque virentis,
> Omnibus incutiens blandum per pectora amorem,

Efficis ut cupide generatim saecla propagent.
Quae quoniam rerum naturam sola gubernan
Nec sine te quicquam dias in luminis oras
Exoritur neque fit laetum neque amabile quicquam,
Te sociam studeo scribendis versibus esse,
Quos ego de rerum natura pangere conor.

All-nourishing Venus, beneath the sliding signs of the sky, everything that lives in the ship-bearing sea and the fruit-bearing earth you yourself fill with life — for through you alone every race of living things is conceived and sees the light of the rising sun. For you, Goddess, the winds leave the cloudy sky and the endlessly varied earth puts forth sweet flowers, the waves of the pacified ocean shine with a soft light. For as soon as the face of spring appears and restoring breezes of the gentle south wind begin to stir, the birds of the air first signal your coming, struck through the heart by your force. Then the wild flocks run over the rich earth, and cross fast-flowing streams, captured by your charm they follow you wild with the desire to procreate. Then through seas and mountains and rapid rivers and the leaf-bearing homes of birds and the rich green fields, all things are pierced with the sweetness of love in their hearts, and through desire propagate their species generation after generation. Since you alone govern the nature of things, nor without you does anything arise within the shores of light, nor is anything lovely or beautiful made, I will endeavor to write these verses to you, in which I will attempt to paint the nature of things.

There is a garden at the edge of the world, in the rim of gold where the sun sets:

Hespera panta pheron osa phainolis eskedas auws
(Hespera brings back all that the bright dawn scattered)

Sappho is said to mean Hespera as Venus here, the evening star, but I have always thought she used the word to mean evening itself, the light thickening to gold on the far horizon.

In the garden of the Hesperides, the children of the evening sky, there are apples made of gold. Though this too is unclear, for *melon*, the Greek word for apple, or, more generally, fruit, is also the word for sheep. Flocks and fruit-bearing trees in the Greek world were gold. They were wealth itself. When Hercules slew the guardian snake in the garden of the Hesperides, did he steal apples or sheep, or clouds dissolving in the sun?

What is an apple? And where did it come from? How can one trace its trajectory west to the orchards of the Seneca Indians, of the spiritualist William Smith, of the New York State Agricultural Experiment Station?

🙰

In another story seeds of gold are carried in a leather pouch across the world as a gift to the poor: corn, the seeds of a hybridized giant grass from an isolated valley in Mexico, traveled north and east across the North American continent, to the Algonquin and the Iroquois, and from them on to Europe, Africa, India, and China in a known and traceable route. The apple was traveling in the opposite direction at the same time, but its path is a mystery.

🙰

The flowering branch appears as an image all along the network of ancient trade routes through China and Persia, in the quick black strokes of early Zen painting, and in the rich, exquisite colors of the ateliers of Safavid Afghanistan, following the lemons, pomegranates, peaches, and pears that traveled west on the Silk Road. The branch was the magic wand: the clone.

The technique of grafting is older than history. It occurs in nature when the wind rubs the branches of different trees together and the raw sap seals the wood. Grafting can reproduce a fruit precisely without the random mess of biological reproduction, which always creates something new. It is a form of miraculous asexual reproduction that has long been associated with monasticism — as in China, where temple complexes kept enclosed orchards and meticulously preserved unusual, sacred trees, like the fossil gingko.

An old friend of mine who has worked in the Sinai desert for years told me that he had seen the nomads there graft domestic apples and pears onto the native drought-resistant hawthorns. They learned the practice from the monks, they said, who carried baskets of earth up to the level inner valley floors of the red granite massif of Mount Sinai, and planted fields of wheat around the hermitages they made out of piled pink stones. The monks planted trees a thousand years ago that remain alive today beside their long-abandoned mountain chapels. The nomads tend the remnants of their hidden orchards high up on the mountain. They practice grafting by wedging the flowering branch of a domestic tree onto a native root stock and packing it in place with mud.

The Roman army grafted apples onto the wild flowering crabs in Britain in much the same way. Might not experienced cultivators like the Iroquois have introduced European apples into western New York with this practice, I asked the pomologist at the Experiment Station, or did they simply grow seedling trees? Champlain wrote of Iroquois apple orchards growing beside the corn along the St. Lawrence in the early 1600s — suggesting that apple seeds were taken up by these experienced cultivators within years of the first European contact.

Were remnant Seneca trees absorbed into the Station's early experimental orchards and hybridized into the multiple commercial American varieties developed by the Station in the twentieth century?

A photograph in *The Apples of New York State*, published by the Station in 1905, shows a battered old Seneca tree on Station land. The photograph was used to illustrate the radical change that the cultivation of apples in America underwent in the nineteenth century, expedited and defined by William Smith's commercial enterprise, and later by the Experiment Station itself: the mass standardization of apple trees by grafting.

An apple tree that grows from a seed is the result of the random pollination of bees bringing pollen from a different tree that the bee was drawn to by its random location and incidental qualities of beauty. Perhaps fellow bees led others to a certain tree, or perhaps a single flower was a new discovery by a single bee from a certain hive. Hence every blossom of every apple tree is fertilized by pollen brought in a minuscule and inadvertent way on the legs of individual, individually driven bees. The process can be broken down to include even the blossoms themselves, which in every detail — the platform on which the bee might be drawn to land, the arrangement of the petals that conceal and expose the genetic material so that it will seduce the insect and accidentally be perpetuated — contribute to this process. The color draws the bees, as does the beauty of the shape and smell, each involved in flagging random bees from the sky. The fertilization of every apple blossom involves a number of different factors, and every fertilization is different, meaning that every seed is a new creation entirely, and the seed that falls from the apple that falls from a certain tree takes a new form as a different tree, a seedling with a randomness that reflects the randomness of all life.

Commercial grafting put an end to this. The grafted fruit is an artificial creation that must be maintained as such by human hands, though the formation of the fruit itself, even on the controlled varieties of grafted trees, remains dependent on the random visits of bees.

Thoreau wrote "Wild Apples" as he watched in alarm the rapidly developing process of mass grafting, of the standardization of the apple. He wrote of apples that taste like themselves, that are always different — that one ate, feeling their bitterness and sweetness at once, in a cold November wind, under the tree itself. "The rows of grafted fruit will never tempt me to wander amid them . . . But I now, alas, speak rather from memory than from recent experience, such ravages have been made!" Of the wild apples — whether the native crab or seedling trees from European cultivars, "I have seen no account of these in the *Fruit and Fruit Trees of America,* though they are more memorable to my taste than the grafted kinds. Apples for grafting appear to have been selected commonly, not so much for their . . . flavor as for their mildness, their size . . . not so much for their beauty. Indeed I have no faith in the selected lists of pomological gentlemen . . . No," Thoreau concludes, "bring me an apple from the tree of life!"

Everyone remembered that Dick Wellington, Dr. Glass's uncle, had taken an interest in the Indian trees. He had formed such a large collection of Seneca artifacts from the Station fields that they became the basis for archeological work there shortly after his death in the mid-1970s. Wellington had come to the Station in the early part of the twentieth century, and became one of the foremost apple breeders in the world, inventing the Cortland, the Macoun, and the Jonagold, among dozens of other varieties. FDR sent his daughter, Anna, to Geneva to work as his assistant in the 1930s, to learn about American apple trees. I remember Wellington at a picnic shortly before he died in Geneva. I thought, This is how I want to be at the end of my life, white-haired, red-cheeked, with a radiant intelligence evident in clear bright eyes.

The pomologist who replaced Wellington at the Station remembered that the day he arrived, in 1949, Wellington had taken him to see one of the old Seneca trees, where "it stood behind a shed on the corner of North Street and the Preemption Road." Would not Wellington, I asked him, who had such an interest in the Indian presence there, have developed grafts from the Seneca trees? And the native American apple — the wild flowering crab, native not only to North America but specifically to the Finger Lakes region — would not Wellington have tried somehow to draw in and perpetuate its qualities?

But no one knew or seemed interested in the answers to these questions about the native trees of the Finger Lakes. The local quality, the regional link, had been lost in the larger picture of the development of a homogeneous salable product.

The pomologist at the Station in the line of Dick Wellington now is Phil Forsline, a remarkably kind man with the patience and generosity of a schoolteacher, who will, if you can catch him, take you through the Station's orchards, now called "the largest apple gene bank in the world," and show you every tree. He has a shelf for the many books that have mentioned his work in recent years. The problem Forsline is trying to solve is a purely practical one: how to expand the gene pool of the commercial grafted apple, the many varieties of which have all been developed from two or three parent strains. The gene pool is so shallow that the entire species would be threatened by the development of a single resistant strain of disease. Hence Forsline, with Barry Juniper at Oxford, has a particular interest in the discovery of the origin of the apple in the Tien Shan Mountains in Kazakhstan by the Russian geneticist Nikolai Vavilov, who published, in 1926, a monograph called "The Problem of the Origin of Fruit Trees."

Vavilov spent years traveling around the world tracking down the native habitats of domesticated plants, and had traced the parent of the cultivated apple to a remnant patch of virgin forest in the Tien Shan. These were descendants of trees that had been protected in the folds of the mountains from Pleistocene glaciation and had been preserved and developed over the ages by the birds that ate their fruit. The seeds of the fallen fruit were carried down from the mountains by wandering bears.

Thus bears were responsible for the development of the apple tree and its sweetness, for the original trees produced fruit that was both sweet and bitter, and bears would eat only the sweetest fruit, and would carry and spread its seeds down from the mountains and west to the plains over millennia.

I was in the Tien Shan Mountains once, the forest from which apples and gold had come, as the accidental guest of an old Kazakh archaeologist named Beken. He had been involved in the discovery of the 2,500-year-old Sakia mummy called "Gold Man," a body found in scarlet clothing, plated all over with gold:

> The entire coat was covered with sewn-on buckles in the form of trefoils and tigers' heads . . . the buckles were cut out of gold foil and sewn or glued onto the red suede as appliqué work. Their rhythmical alternation against the red background of the coat and thigh boots created an openwork effect. The headgear was decorated with buckles in the shape of leopards, tigers, sculpted figures of horned and winged horses, birds, an ibex on the top of the hat, and plaques depicting a mountain with the "tree of life," others in the form of birds' wings.
>
> (K. Akishev, *The Ancient Gold of Kazakhstan*)

Here was the tree of life on the edge of the Taklamakan Desert, in the green lower reaches of snow-capped mountains, above dry gravel courses of desert riverbeds lined with wild irises, and oases of white-trunked poplar trees and tiny round-topped Chinese elms filled with lilac rollers and kingfishers the color of lapis lazuli.

The landscape was an illustration of the mixing of Asian and Caucasian blood that has gone on in the steppes since prehistory. We were looking for rock drawings in the red sandstone mountains and had come down into the Ili Valley in China where the script was Arabic, the language Turkic, and the architecture Russian. In Urumchi there were tall fair Scythian mummies with blue tattooed cheeks.

Beken was looking for Scythian burial mounds, and lived out of a converted Soviet Army bus. It was the summer of 1991. We did not know that the Soviet Union was about to fall apart. Nothing had happened, yet there was an odd mood of laxness, almost of gaiety, in the air. When the bus was stopped at a military checkpoint and the soldiers demanded, "What have you got in there?" the men said, "American women from New York," and the soldiers winked and waved us on.

Flocks of sheep spilled over the steep green sides of the mountains. There were no roads. Occasionally we would see a lone black bull (*"Buik lubit krasnia,"* Beken would say, laughing at my old red hooded sweatshirt. "Bulls love red.") In the evening Beken fished for tiny pink-flecked trout in the mountain streams, which looked as if they were made of beaten gold.

We sat in the dark after the supper fire was gone, naming the stars as each of us knew them, for we had no common language. Beken, an old Asian with a white beard, looked to me for all the world like an Eskimo. He remembered the names in Kazakh, the Turkic of his nomadic childhood. Stalin starved the Kazakh nomads and forced those who remained into the urban life of the capital, Alma Ata (which in Kazakh means Father of the Apple), where they were forced to speak Russian, the language of the state. Beken became an archaeologist in order to take occasional refuge in the hills.

"Eto archaeologie!" he would say. "Now this is archaeology!"

6

BEES

THERE IS A KIND OF REST that is not sleep exactly. It covers the mind like the film that covers the eye of an animal at night — the translucent second lid. Light penetrates, thoughts float freely through levels of the mind, but the body is at rest. After nights of lying outside in this condition, one's eyes, opening fitfully, begin to fix steadily on the changing patterns of stars. To someone who lives outside, these patterns are not mere talismans of lost belief. They are a map defining the horizon, as it slips with each day and with one's progress.

The concepts of time and the sky have only diverged as the brightness of each settled place blots out and separates it from whatever worlds there are beyond, both earthly and celestial. Our very word for hour, the root for time and season, was originally used to represent the precincts of the sky. When Sappho, watching the moon and Pleiades set, writes, *mesai de nuktes, para d'erket' ora* "It is the middle of the night, the hours go by," she does not refer to time abstractly passing, but to the constellations that pass overhead as she watches. Expressing the same simple visual truth, Herodotus says of traveling east or west, "on towards evening, on towards morning," picked up by Barrie for the flights of Peter Pan.

Of all the different hours of day I have loved most the last hour before dawn, the last frieze of stars when the horn of the sun meets the horizon on the celestial equator and stops the zodiac like a dial.

That the zodiac were signs in the pictorial-alphabetic sense was a revelation to me. The first time I looked up and recognized Aquarius — not with a chart, but by the silver wavelets, the drops that fell around it (and seemed to float, and perfectly to represent the universal ancient sign for water, a single wave, which ultimately works its way into our alphabet as the letter *m*) — and Scorpio by its bright red heart, they became my toys from childhood, present everywhere without having to be brought, and I could feel through their grip on my imagination how pervasive their influence had been.

In the mid-1980s I took a tiny ground-floor back apartment on Claremont Avenue, off Riverside Drive, in New York. It was quiet and dark, and I felt like a ghost in the old familiar neighborhood. I went sometimes for long stretches without talking to anyone at all. The hours were marked by bells from Riverside Church, which rose up bone white and sharp against the blue winter sky outside.

When I arrived the apartment had the fresh chemical smell of new paint. I remember the desolate first night — the snow falling silently, heavily against the blackness of the bare window glass. I brought with me two shopping bags packed with things from my room in Geneva — my grandmother's silver tea set from Ontario, a small faded Kazakh rug, Bedouin silks, a translucent blue willow Japanese teacup that had belonged to my grandfather in Albany, small white cardboard boxes lined with cotton that contained familiar fossils so handled and known that they were almost toys: a sectioned rugose coral like a tooth or growth of horn, lined evenly within as though with spider webs hardened into stone; a perfectly rounded trilobite punched out of the chipped slate on our shore at Seneca Lake.

I covered the windowsill with lichen-patched twigs and bracket fungus and a kind of rich green moss from Riverside Park. Every morning I would pour a pitcher of water on the sill. The moss was so thick that the water never fell to the floor.

I was preoccupied with one question in those days: how to get back to Egypt. But I loved the apartment on Claremont Avenue. It gave me back Riverside Park, where I knew every tree. I watched two European birches die that year at 108th Street, and a big-toothed aspen at 115th. I used to walk under the elms in the early morning and cut fresh branches of flowers and leaves, which I hid in a sack from the street of the canvas makers in Cairo.

My desk on Claremont Avenue was a black card table in a corner between two windows. I had a folding chair, lent by a friend. My bed was a mattress on the floor. Outside one window was the broad platform of a fire escape beneath a huge old gingko tree, lime green in spring and golden in the fall. In the afternoon I would push back the heavy accordion grate, having unlocked its rusted padlock, and climb out onto the fire escape with the little faded rug, the pillows from my bed, and a tray of tea.

I sometimes went to see a friend who worked at night at the Columbia telescope on the roof of a nearby building on 120th Street. The telescope was a century old, not much in use, and Joe was there at night as a sort of guard from the astronomy department. He was a skeletal man, long and loopy, like a swamp dweller in a fairy tale. He showed me Halley's comet that year and Saturn's rings, set as a circular rainbow, radiant as a jewel, perfectly formed against the black night sky. The old wood floor would turn and creak, the tongue of copper with its lime green film would slide back, as the telescope swung out into the night. I stood on a stool, as Joe showed me how to squint to see out through the immense tubular eye.

I used to come into New York on the Hudson train from Syracuse and walk across Grand Central Station with two black garbage bags. In one were four or five frozen trout, and in the other, a honeycomb in a drawer-sized wood frame I had taken from the honey house.

Whenever I went up to the Finger Lakes I would go to the honey house on Kime Road just off the northeast curve of Seneca Lake, where Bob Kime in his rumpled bee suit, flecked with drowsy disoriented bees, would show me his recent take. We could see, even in a single frame, the tremendous variety of flowers from trees and fields the bees had hit over the weeks past. Throughout each sheet of comb the cells were variegated with honeys of different densities in patches of pale gold deepening to amber, for honeybees will work a certain crop of flower until it is gone before moving on to the next.

Bob and I would dig our fingers into the combs and try to identify the different tastes, and thus the progress of orchards and fields the bees had traveled that month. Bob had worked for years in the lab at the Experiment Station and he knew a great deal about flowers and fruit and their qualities and smells. He was also the son of one of the old farming families on Seneca Lake, and he knew a great deal more than a researcher would about where things were and how they worked. So there we were, tasting right out of the hive the honey that was the essence of the fruit.

One year we found raspberry that was crystal in the comb, and once a dense wild plum that was so strong it was almost intoxicating. At first we couldn't figure out what it was. We would uncap the combs with the hot electric knife, which moved stiffly and was hard to hold. There was the wonderful smell of wax burning as the comb gave, the wax falling away into a tub to drain and be sold separately. Then we would fit the uncapped frames into the extractor and spin those rare flavors together into something more like what one would expect honey to be.

I liked to have my honey straight out of the comb, and Bob gave me a frame to take back to New York. I boiled up the wax on the gas stove and made the apartment smell like the honey house. I sat on the fire escape under the gingko tree with a plate of Kimey's honey and ate it with a spoon. I couldn't keep bees in New York, but I knew every hive of wild bees in Riverside Park. There was a hive high up in a tulip poplar just south of 110th Street, and another near the ground at 121st Street in an old locust where the trunk had been split vertically with an ax.

Bob suggested I take a job teaching beekeeping in Darfur for the Ford Foundation. He heard they were looking for someone, and he knew I was looking for a way to go back to Egypt and Sudan. I thought of Riverside Drive as the wood between the worlds that summer, the safe neutral forest where one is lulled to sleep beneath the trees.

I would get up in the morning and walk over to Morningside Heights to watch the sun rise. Then I would walk down through Riverside Park to see what was out. Sometimes I saw remarkable things: a question-mark butterfly on an ailanthus leaf in a thunderstorm, its ragged wings dusted over with the purple of the sky; a brilliant orange mass of jack-o'-lantern toadstools that came up all at once after rain; a shelf of pure white angel wing that filled the soft trunk of a dying elm.

I remember the successive waves of flowers — the corridor of basswood down the hill south of 103rd, the wild roses up near Riverside Church, the old honey locusts ringed with thorns at 96th, their waves of heady smells dividing the weeks of June, each crop of flowers rapidly filling with bees.

My old friend Dick Miller would come through New York on his way to and from North Yemen. He would take me down to Rockefeller Center and show me the letters etched in gold about citizenship and sociability, the joys of city life. He would fix me in his cornflower blue eyes — the pupil of one eye ripped at the base, bleeding in a jagged black streak through the pale bluish rim of the iris — a prismatic eye, I thought — and wave his arm toward the crowds, the lighted buildings, and say, "Well, you won't see anything like this in Darfur."

In the end I did not go to Darfur to teach beekeeping for the Ford Foundation. Though one day on an oasis in the Libyan desert I

heard what I thought was the sound of rain and found myself standing in a bee yard.

In 1910 an ophthalmologist named C. von Hess asserted that fish and
all invertebrates were colorblind.

"But I could not believe it," wrote the biologist Karl von Frisch.
"It was easier to believe that a scientist had come to a false conclusion
than that nature had made an absurd mistake."

In 1910 von Frisch began a series of basic empirical experiments
with squares of colored cardboard, sugar water, and sheets of cello-
phane. Over decades of close observation, he came to understand the
primary senses of insects, and ultimately cracked the language of
bees. In 1973 von Frisch won the Nobel Prize.

Von Frisch saw everywhere the organic significance of color, beginning with the color blue, the blueness of the sky. The sky is blue
when the light of the sun is scattered and diffused by the high small
particles of the atmosphere. In the blue sky the sun's light vibrates
away from the particles of air at a definite angle, like laundry hanging
in the wind. The human eye cannot detect this angling of light; it
works like a camera, reproducing the image of what is before it in an
approximate condensed way, creating an illusion of solidity, of fixity.
But a honeybee sees the blue sky as shifting patterns of polarized
light. A honeybee can read the position of the sun at any time by seeing a patch of blue sky.

The eye of the bee is made up of thousands of tiny light recep-
tors, each eight-petaled like a lotus or a rosette, with the light piercing
through its center. The perception of light and color differs slightly in
each receptor, so that the overall visual effect is multifaceted, frag-
mented.

The angled light of the sun to such an eye is a map. The honeybee can reproduce that map by dancing, dancing the angles of light.
The dancing is contagious. The bees near her learn the dance, dance

with her over the combs, and then can read and follow her path through the sky.

Bees see in the blue range, the short end of the color spectrum. A bee sees blues a human being cannot see. To a bee, ultraviolet, the ultraviolet rays reflected back from the petals of a flower, is a vivid color, red is black, and green is gray. Hence (von Frisch concludes) there are few scarlet wildflowers in Europe, no cardinal flowers, for in Europe there are no hummingbirds.

To a bee, a newly opened flower arises from its nest of dull leaves as a vibrant mass of color. The form and fragrance of flowers are not accidental. The yellow streak on the inside of an iris is an arrow, leading the bee in to the hidden pool of nectar at its base.

The iris supplies the nectar for the bee. *Nek tar,* "that which overcomes death." As the bee plunges down to the nectar, unfurling her long, red, strawlike tongue, her fur is dusted over with pollen, the protein-rich male seed of the flower. As the bee moves on she inevitably carries the pollen to the female parts of a neighboring flower of the same species, and so enables the fruit to form, the seed to grow, the flower or tree to continue.

"Consider in their entirety," von Frisch wrote in 1971, "the accomplishments of these small insects, the bees. The more deeply one probes here the greater his sense of wonder, and this may perhaps restore to some that reverence for the creative forces of nature which has unfortunately been lost."

Honeybees and flowering plants depend on each other and evolved simultaneously, suddenly, sixty million years ago at the end of the Cretaceous, the age of dinosaurs. Beekeeping is sometimes called the earliest form of agriculture. A rock drawing from the Mesolithic in a rock shelter in Spain shows a person reaching into a tree for the honey inside, surrounded by thick black bees. In the other hand the person holds a small vessel, a honey jar.

One can only imagine the impact of sweetness in the millennia before the refinement of cane sugar: sweetness as something strange, an intoxicant. Hence one can imagine the value of honey: a treasure taken at some cost of pain.

Then there is the insistent symmetry, the fragile, insistent symmetry of the honeycomb, and of the lives of the bees themselves.

❧

The native crops and flowering trees of the Finger Lakes, like *Crataegus,* the hawthorn, and *Pyrus coronaria,* the native apple or flowering crab, were pollinated by a vast range of pollen bees native to the American continent, like the orchard bee and the bumblebee. These are not hive bees and do not produce any quantity of honey; the Iroquois sweetener before the introduction of the honeybee was maple sap. The populations of native bees have dwindled dramatically because of habitat destruction and the use of pesticides. Reliable crop pollination can now be accomplished only by carefully monitored hives of honeybees. This kind of controlled pollination has been around for a long time. In pharaonic Egypt beekeepers moved hives seasonally up the Nile on rafts to pollinate the fields and make honey — a double boon. But now it is critical.

❧

Bob had an interest in honey that was as thorough as his interest in anything he took up. One day he stirred a teaspoon of honey into his tea. He saw the sediment drift to the bottom of the cup and realized that honey worked as a clarifying agent and could be used as a preservative. He patented the use of honey as a preservative in fruit juice and wine, eliminating the need for artificial chemical agents. (His method is used around the world now and has been cited as a classic example of empirical science.) Not willing to waste anything, he used the beeswax left over from his hives after the fall honey harvest to make a skin cream by melting the wax with glycerine on his kitchen stove. The cream was used by the workers at a nearby cement factory to heal sores on their skin. Bob got into beekeeping because the Station needed a beekeeper to assure the pollination of its orchards;

the former beekeeper had left after being stung forty times in a single day.

By 1981 Bob had almost two hundred hives in bee yards around the Finger Lakes, and many farmers depended on him for the pollination of their orchards.

🐝

Beekeeping is seasonal work. The bulk of it takes place in the fall: the honey harvest. In my mind the honey harvest is mixed with red leaves and a bite in the air, the last of the asters scraggly and edged with brown.

Bob and I would head out at seven in the morning in his pickup, eating apples from his mother's orchard as we drove along, down the sweeping farm roads around the lakes, down roads of matted grass that ran alongside fields, and into brush clearings of sumac and elder where his bee yards were hidden away.

He kept us going through the long days by telling hunting stories in his clear reedy voice, his steadying voice:

"A male turkey will never go to a female turkey, see. And every male turkey has a harem of four or five hens. So to get a male turkey you have to make him think that there's a female out there somewhere, and that it's worth his while to go to her. So you have to be downhill . . . 'cause he'll only go downhill, it's so little effort . . . in a tree . . . before dawn, when he's just waking up . . . and then . . . I mean, most people don't get more than one or two turkeys in a lifetime. It's that difficult."

🐝

All the while Bob's attention is completely on the hives: he moves through each yard in a concentrated, methodical way. We begin by taking the bee suits out of the back of the truck and putting them on over our jeans. The suits are a heavy white cotton, safety-pinned together where they are torn. We stuff the pants legs into our work boots and zip on the hoods that cover our heads but leave a mesh opening for our faces — all only enough to discourage a bee, really. A few usually get in around the wrists and knees. Bob always gets stung a few times.

Bob explains what is going on inside the hive, where I see only a mass of bees. A hive is like a city, he would say, with the queen at its heart. Most bees hold a progression of different jobs in the hive and go through a series of physical transformations to meet each job.

One day in a bee's life is comparable to years in a human life. The life span of a worker bee is four to six weeks, while the queen will live up to four or five years. Von Frisch in his experiments numbered every bee with a dab of colored paint. He monitored the bee's career throughout its life. He saw that each honeybee acted alone, and had its idiosyncrasies. Yet the population of a hive functioned almost as particles of a single mind.

"You cannot keep just one bee," von Frisch wrote. "This is not as simple as it sounds."

A hive in summer contains thirty thousand bees or more — a good-sized city. In the spring and summer a queen lays up to fifteen hundred eggs a day — at least one egg a minute day and night. As the infant bees emerge, just as many older workers die, having shredded their fragile wings to pieces in foraging flights.

All a beekeeper can do is offer the bees an artfully arranged substitute for a hive in a hollow tree — wooden boxes stacked up with a hardboard lid. In the controlled, artificial hive a metal grid called a "queen excluder" is laid on top of the second box to prevent the queen from climbing up into the supers — the boxes on top of the stacks — and filling them with eggs. The pure supply of honey in the supers is what the beekeeper is after.

Worker bees easily pass through the queen excluder, moving freely up and down to perform their tasks. The entrance to the hive is at the very bottom of the lowest box.

When we open the hive we rarely see the queen. She lives where the young are hatched and reared, in the middle of the lower combs, surrounded by pollen. The queen wanders through her territory all day long, slowly, regally, surrounded by her changing court of worker bees, which protect and wash her, brushing back her hair, bringing her food, carrying away her excrement. They touch her all the while with their antennae.

As they lick and touch the queen the workers pick up substances secreted by her body. The workers constantly touch one another, and in doing so spread her pheromones throughout the hive, signaling that the queen is alive and well.

The workers create a thin layer of larger cells on the edge of the comb for unfertilized eggs, which the queen lays at will. The unfertilized eggs develop into male bees, drones, larger than the worker bees, even burly. The drones do no work. They wander through the hive as they wish, taking pollen and honey, and making messes wherever they go.

A few thousand drones are produced in the spring and summer. Their sole function is to mate with a queen. The worker bees seem to grow increasingly impatient with the behavior of the drones. After a few months they begin to pinch and push them, and at last they literally drag them out of the hive. The drones cower together and try to hide in the empty cells or in the space between the combs and the inside wall of the hive, but ultimately they are found and driven out, nearly every one.

Most of the worker bees are dead by the end of winter, and the population of the hive has dwindled to around six thousand. The remaining bees cluster together in a ball around the queen. The outer layer of bees in the cluster generates warmth by spasmodically moving their flight muscles back and forth. The outer layer of bees changes continually, as the workers become exhausted and crawl to the stillness and warmth of the inside of the cluster. At the winter solstice the surviving worker bees raise the temperature in the hive to 93 degrees Fahrenheit, the necessary temperature for the creation of young, and the queen begins to lay eggs, bringing the hive back to life again.

The intricate stages of a honeybee's career:

Maggotlike larvae hatch from the eggs laid by the queen. Young workers five days old or more who have been feeding heavily on pollen produce for the larvae a substance called bee milk or royal jelly. Rich in protein, acidic and bitter, this milky white cream is fed to the larvae and to the queen throughout her life.

The larvae grow rapidly to five hundred times their original weight. In six days, when their bodies have swollen to fill their cells completely, they spin silk cocoons about themselves and begin to pupate. Older workers cap the cells with a covering of fine, light wax. The new bee miraculously evolves, translucent and perfect within the pupa, and on the twelfth day after the capping chews her way out through the lid of her cell.

For the first three days of her new life the young bee walks around the brood comb, licking clean the newly emptied cells and polishing them with her saliva. Only if a cell has been prepared in this way will the queen lower her long, slender abdomen into it and lay an egg.

The young bee at first spends her time sitting around, combing her hair ("just like a teenager," Bob says). After three days the hypopharyngeal glands in her head have fully developed and she becomes a nurse. She begins by feeding the older, bigger larvae, but as she becomes more skilled she feeds the younger, more delicate larvae with the bee milk from her mouth.

At the end of ten days the glands on the head of the nurse begin to shrink; she can no longer produce royal jelly. The four pairs of wax glands on her lower abdomen have now reached their full size, and she begins to secrete wax. The wax forms in scales and slides out from the underside of her abdomen. The bee retreats to a warm, crowded part of the hive, where other bees her age hang suspended.

In the warmer months, when the population of the hive swells, the wax-producing bees begin to build layers of new comb. They gorge themselves on honey and in a few hours secrete immense quantities of wax (consuming six to eight pounds of honey to produce one pound of wax). Workers roll the wax up away from their abdomens with their legs, and chew and soften it with saliva. With their mouths and feet they manipulate the wax to form the mathematically precise sheets of fragile hexagonal cells that honeybees have routinely created for millions of years.

In the second stage of life (between ten and twenty days old) worker bees go to the entrance of the hive and receive in their mouths the nectar that older foragers have brought back in their honey stomachs. A forager bee may visit hundreds of sources of nectar (for example, each of the florets on a head of clover) to fill her honey stomach — a crop, a kind of holding tank — just once. Every teaspoon of honey may require thousands of trips to the field by forager bees.

The receiving bee holds the nectar in her honey stomach as she carries it up to the cells where the honey is stored. When she disgorges the nectar she adds to it fluids secreted from her salivary and now contracted hypopharyngeal glands, filled with enzymes to purify and preserve the honey.

The receiving bees pack the nectar into honey cells. Other workers stand above, thrumming their wings to evaporate excess moisture and thicken the nectar. When a honey cell is filled, a wax-producing bee caps it over with a vaulted lid, and the altered nectar is left to age.

The bees now begin to make their first excursions outside, short flights to familiarize themselves with the territory. Some are responsible for clearing refuse from the hive; they carry the corpses of other bees a distance away from the hive and drop them to the ground. Others stand by the entrance and guard it from intruders, including honey-stealing bees from other hives. The guards run their antennae over every creature that attempts to enter the hive, and attack intruders with their stingers. The stinger of a worker bee is a modified ovipositor, and is used primarily for combat with bees and other insects, like a sword. When a honeybee stings a large animal, such as a human being, the stinger remains lodged in the thick skin of the victim and the bee disembowels herself trying to pull it free.

In the final period of a worker bee's life, from about the twentieth day until her death, she becomes a forager, collecting water, resin, nectar, or pollen. Her body is ingeniously shaped for the work. Her inner back legs are covered with bristly brushes that mesh together on opposite legs. The bee brushes away the pollen from a flower's thickly coated stamens, then forces the pollen through the stiff hairs of one leg with the bristles of the leg opposite, as though combing or sifting it. On the outer side of her back legs are deep indentations called pollen baskets. She moistens the pollen with honey that she has brought from the hive in her honey stomach. She forms the pollen into a ball, like bread dough, packs it into her pollen baskets, then flies back to the hive with bright bulbs of pollen bulging from her legs.

A forager also fills her honey stomach with water. Water is used in the hive primarily as a cooling agent. In summer the bees spread a fine film of water over the comb and stand above, thousands of them at a time, thrumming their wings to evaporate the water and bring the temperature of the hive down. The workers see that the hive always remains at the temperature necessary to sustain life, never too hot or too cold.

The queen bee develops from a normal worker egg, laid in a specially constructed, large, acorn-shaped cell, and is fed from birth huge quantities of food. A hive in brood season always has a few developing queen larvae. If the population of a hive becomes too large, the colony splits and the old queen flies away to found a new colony with a swarm of worker bees. This leaves the original colony without a queen. But in a few days a new queen emerges, in a brutal drama of succession that might have been described by Frazer in *The Golden Bough*. The first queen to come out of her cell walks over the comb and makes a shrill piping noise by crouching and vibrating her wings. She is warning the other queens that she is abroad. They respond with an answering vibration. If another queen emerges from her cell the first queen fights her to the death with her stinger. Then the surviving young queen goes through all the queen cells, bites through their lids of wax, and kills the potential rivals within.

Now it is time for the new queen's nuptial flight, the only time she will ever leave the hive. The workers prepare her, shake her bodily, and then push her out of the hive. She ventures some distance away, and her scent entices males to follow.

The Dadants' *The Hive and the Honeybee* describes the mating of the queen: "Reports mention a comet-shaped swarm of swiftly flying drones weaving hither and thither, presumably chasing a virgin queen which was supposed to be at its apex. Observers have reported hearing a sharp crack at the time that they believed that copulation occurred and that the queen broke away from the drone which fell dead or dying to the ground."

On her bridal flight the queen mates with ten or twelve drones. She has a sac in her abdomen, a spermatheca, in which she stores the collected sperm for the duration of her life and from which she fertilizes her own eggs. The mix of sperm from the drones of different hives provides the genetic diversity necessary to maintain the health of the colony, for a colony of thirty thousand honeybees has only one mother.

🐝

Bob and I walk up to the hives in our bee suits. The curious guard bees come out to explore, dancing before the mesh around our faces. They land on it and relax a little, combing back the blond fur on their backs with their long red tongues. With our hive tools we crack the propolis sealing that holds the lid of a hive in place. Propolis is a material the bees make from plant sap and pine resin. It is filled with antibacterial agents and drives insect pests away. In a natural hive honeybees paint the whole inside of a hollow tree with propolis, as though to sterilize it, before building the comb within. Bees use propolis to cover up disagreeable foreign agents, like beetles, that are too large to carry out of the hive. In winter propolis is hard and brittle. In summer it is warm and sticky. To the beekeeper it is a pain in the neck. And there is a sense of pleasure in feeling it crack. We separate the frames. Now we are in.

Immediately we hold up Bob's old tin smoker, which is filled with twine soaked in kerosene. We light the twine with a match, and as it smolders we blow the smoke up with the smoker's bellows and funnel it in among the frames. The bees think the hive is on fire and start filling their honey stomachs to carry the precious honey out of the burning hive. Soon the heaviness in their bodies makes them sleepy, and they become too confused to attack us.

Bob tells me that in Africa they set fire to whole trees to get at the honey. "That's why," he says, "they call them African killer bees — they're just so mad." We are always joking about the "African killer bee," *Apis mellifera scutellata*. In 1956, a few scutellata queens were imported from South Africa to Brazil in an attempt to improve the local stock of honeybees. The African killer bees have been moving steadily north ever since at a rate of one to two hundred miles a year; they have recently been seen in Texas, California, Arizona, and Nevada.

The scutellata are easily prompted to sting, and do so in large numbers. A friend of mine who was in the Peace Corps in Zaire once jumped into a river to save his life after he accidentally walked by one of their hives. Within moments, he said, the bees had completely covered him.

One day Bob was up on Bean's Hill, high above Geneva with views of the countryside all around. He was carrying a truckload of hives with his son Shawn. Bob's nephew Scott was in another truck with another load of hives. They passed through a construction site where the road was rough and full of potholes. The hives in Scott's truck were shaken loose, and as he turned a corner four hives, which were bound together by a rope, fell off the truck and were smashed open. Two hundred thousand angry bees flew up into the air. Scott was no longer in his bee suit but, without stopping to think, jumped out of the truck to see what had happened. The bees plastered him at once. He screamed in pain and jumped face-first into a ditch with the bees covering him. Bob was struggling to get back in his bee suit in the cab of the truck and get to Scott. Bees were pouring into the windows of cars all along the road, through the doors of banks and restaurants and grocery stores, where people were trapped for hours that afternoon, waiting for the masses of bees outside to settle down. A man in a freezer truck grabbed Scott and threw him in the back. Within a minute the bees fell off in the cold. Scott's eyes were swollen shut, his tongue was so swollen he could not talk, and he was rushed to the hospital. Had he been left much longer to the mercy of the angry bees he could easily have died.

🐝

Honeybees came to the American continent with European colonists as early as the sixteenth century. They came in skeps, upside-down baskets made of woven straw or twigs.

The traditional method of beekeeping was to allow a colony of bees to fill a skep with honey, then to force them out into an empty hive, or sometimes to kill them by filling the skep with sulfur smoke, or dropping it into boiling water.

In 1851 an American clergyman in Philadelphia named L. L. Langstroth discovered that honeybees will move around between the inside edges of a hive and the combs themselves in a space that is between a quarter and three eighths of an inch wide. This created the possibility of movable, hence removable, frames. Individual frames matted over with sheets of comb could be fitted, suspended into boxes, and lifted out without damaging the hive. The frames could be replaced, they could be periodically checked, and the beekeeper could monitor the life of the hive and remove the excess honey that is produced every year without harming or displacing the bees. Langstroth's "bee space" and removable frames revolutionized bee-

keeping, and allowed for the first time the in-depth study of the lives
of bees.

In the large supers Bob uses there are nine frames in a box. We
pull the frames up one by one, brushing off the disabled bees with a
soft brush, and examine the combs. A hive ordinarily produces fifty
to eighty pounds of honey in a season and can make up to four hundred pounds if conditions are good. Bob checks the cappings on the
combs — thin, slightly translucent swirlings of white over the honey
when the combs are healthy. He judges by the heaviness of the frames,
and of the supers themselves, how much honey he can safely take off.
If a super is filled with good, honey-packed frames, he takes the
whole box.

If half the frames are full, Bob takes a portion of them, moving
the frames into an empty super at his feet. When the super is full, we
load it into the back of the truck. It is very heavy work. After a while
we feel like we are standing on a sidewalk in scuba gear. "Hottest
work in farming," Bob says. We cannot, of course, touch our faces
through our masks, or our hands through our thick gloves.

Bob goes around with a bag of white sugar. He leaves a heap
inside the hive lid wherever we have taken supers away. The bees
convert the sugar into honey to replace what we have taken. Bob is always careful to leave enough honey for the bees to get through the
winter, and says that it is heartbreaking to come back in the spring
and find that a hive has starved to death, the cluster of bees all stiff
and dry.

In the afternoon we take a break and go to Sweet Sue's for a beer,
"just to scare everyone." We are still in our bee suits, and bees swirl
around us here and there.

At the end of the day Bob sometimes says, "Why does anyone go
into this? That's what I want to know. I mean, it's such hard work.
And I'll tell you why. It's in your blood. It gets in your blood."

Ten years later our bee talk was about the varroa mite, which had wiped out virtually every hive of wild bees in the United States in a decade, and devastated the American beekeeping industry.

"It makes you sick," Bob said, telling me about the time in 1995 when he first opened one of his hives and found the young bees half eaten away, yet still alive, the older workers struggling to drag the crippled bees out of the infested combs and destroy them. The varroa mite is not much smaller than a deer tick. A beekeeper can see in the hive the mites attached to the bodies of the bees, living on their blood.

A female mite rides around on a worker bee in the brood comb. Just as the worker is about to cap a cell in which a larva is pupating, the mite jumps in. Inside the sealed cell the mite lays her eggs. The eggs hatch and the young mites feed on the trapped, developing bee, first chewing off its legs and wings.

The varroa plague has been compounded by the simultaneous appearance of a tracheal mite. The tracheal mite gets into the lungs of bees, breeds there, and weakens the respiratory systems of bees so that they do not have the strength to fly.

Both mite populations were first detected in the mid-1980s. The mites came into the United States through Ohio, it seems, on an illegally imported queen. They then spread rapidly across the country with the migrant beekeepers who move from the South to the North every year, from Florida up through the Carolinas, through New York to Maine, with the waves of flower and fruit crops, from lemons and oranges to apples and pears to cranberries and blueberries.

"Sometimes you open a hive after the winter and the queen is there with ten bees," Bob said. "Imagine eighty percent of your hives are dead. That's what happened to everybody except me. You know me. I took it seriously right away."

A chemical insecticide called Apistan had been developed to kill the varroa mites, but they were rapidly becoming resistant to it. Researchers at Cornell were hoping to extract an herbal remedy that could be used in a hive. The smell of sage or rosemary or thyme is the opposite of the nectar or perfume of a flower. It is an insect repellent that the plant itself has developed to repel predatory insects, and it has been seen to repel varroa mites.

For tracheal mites the one effective chemical is menthol. "Menthol is a cool burn," Bob said, "like a cough drop. You think it's cooling your throat. But it's actually burning you. I hate menthol now that I've started using it in the hives."

A tremendous amount of care and attention has to go into beekeeping now just to get through the year. Many beekeepers have lost all their hives to the mites and have left the business.

"Now at least people might begin to recognize how important beekeeping is," Bob said. Beekeeping is the most important and the most silent branch of agriculture. Eighty-five percent of the fruit industry depends upon it completely, as does the nursery business. Bees pollinate alfalfa and clover fields, the food of cows, as well as trees and wild flowers.

"Beekeeping is a tiny one-hundred-million-dollar industry," Bob said, "that creates fifteen billion dollars' worth of food."

Pollination by honeybees is so pervasive, and so little acknowledged, that people don't realize that their gardens are thinning, their crops are failing, because of the sudden absence of bees. Now an orchard may only reliably produce if a beekeeper is hired to put hives in that orchard and monitor them carefully through the year. A farmer cannot rely on the haphazard visits of wild bees or bees from distant hives anymore.

The problem is essentially the virtue of beekeeping. Beekeeping cannot be industrialized. Bob said, "You own the beehive, but you don't own the bees." You cannot force a colony of bees to carry on its complicated work, you can only observe, and help where you can. This involves tremendous patience, the kind of patience that, as a recent article in *Bee Culture* magazine observed, is "no longer amenable to the modern attention span."

One cannot imagine bees and honey going out of the world, for the fascination with bees is as old as humankind. Honey and bees figure in the earliest representations, in remnants of the earliest written language, in the Sumerian "goblets filled with the blood of trees," the food of the immortals — frankincense, honey, and myrrh.

Bees dwelt in the tree of life, the sacred grove, the enchanted forest, the timeless realm between the living and the dead. The sweetness of honey had a mystical sense — it was a taste of the sweetness of wisdom, the wisdom of the numinous other world.

The risen Christ eats of a honeycomb. Apollo was raised in the grove of Parnassus by the Thriae, the Bee Maidens, who taught him to prophesy. The name Deborah in the book of Judges is the Semitic word for bee, meaning prophetess. Aldebaran, the red star that is the eye of Taurus, is "the bee" — red, like Antares in Scorpio, suggesting its quality of pain, its sting. Praesepe, the Beehive, a star cluster in Cancer that seems to vanish and reappear, is a crack through which souls enter and depart the material world. Plato in the *Ion* describes the *sophos* as a bee, an instrument of the divine, having prophetic powers.

Bees were everywhere an essential part of the depiction of wild nature. The priestesses of Artemis, the goddess of the mountains and the forest and the moon, were called Melissonomoi, "beekeepers." A beautifully mysterious fragment of Alcman describes the creatures of the wilderness:

> The peaks and the valleys of the mountains are asleep
> The headlands and the bays
> The tribes of every creeping thing the black earth feeds
> The mountain-going beasts and race of bees
> The monsters in the depths of the porphyry sea
> And the slender-winged birds are asleep.

An ancient Egyptian story tells how bees were born from a tear in the eye of the sun, touching upon their goldenness, the sense that honey is gold, materialized sunlight.

Samson (from *shams*, the Semitic word for sun) comes upon a lion, a creature of the sun, a creature of gold. He tears the lion apart with his bare hands. Later he finds the lion's carcass filled with bees. He reaches inside and takes some honey in his hand and eats it. "Out of the eater came what is eaten," says Samson, "and out of the strong came what is sweet" — the painful rending of strength, the hard, reveals within it what is soft, the sweetness.

One dimension in the golden quality of bees and honey represents thought itself, the stored thought that is all of human accomplishment.

The *kellos* of the honeycomb was the cell of the monk. A Coptic monk in the Egyptian desert once said to me, "You pray and pray until the prayer becomes like honey in your mouth."

The last time I saw Bob I was visiting my parents in Geneva and went to track him down. I found him in his lab around midnight. I hadn't been there in years and felt a chill at the strangeness of the sight, as though I were seeing it again through the eyes of a child. Through the wall of windows thirty feet high, segmented in panes of old glass, the dimly illuminated copper-green equipment of food processing stood poised like giant insects in the dark.

Inside Bob was working late, as usual, intent and beelike in his industriousness. On that particular night, he was bottling berry wine he had just made to distribute at the New York State Berry Growers Association conference in Syracuse the following morning.

I followed him as he walked around with a clipboard, holding up to the light small jewel-like bottles of raspberry, strawberry, peach, and plum wine, grading them by their color (raspberry in five varieties or grades, strawberry seven, etc.) and pouring out shot glasses for us to taste as we walked along, Bob in jeans and baseball cap, his hair gray now beneath it, laughing as usual, saying in his dear, familiar reedy voice, "It's easy. You just take thirty pounds of fruit, fifteen pounds of sugar, two gallons of water, add a little yeast . . ." He handed me a plastic bubbler so I could try it at home.

"Yeah, it's easy for him." His assistant, Tracy, laughed.

We were getting a little woozy with the wine. It was delicious, like sweet perfectly ripe fruit, without a trace of a foreign or chemical substance.

"Those poor grape wine snobs," Bob said, "they don't know what they're missing."

7

THE SILVER FOREST

THE SILVER LIGHT has retreated onto the branch tips, silvering now only the shells of the trees. It is November. The air is cold, and there is still a burning color to things. The red squirrels are whistling in the maples by the road. Last night, as we came in, a barred owl flew raggedly over the hood of the car in the dark, shuddering slightly in flight like a giant moth. We stood beside the house in the starless night. From within the line of tall red pines a hundred yards away we heard a sudden sound — a scream, shrill and prolonged. The pure sound itself made us stiffen with fear, for it was an almost human scream. The night was abruptly still. And then another sound began, in pitch like the first, but rising, winding and unwinding into many voices, low and high, that broke at last into sharp short cries.

Where we live in this upland valley east of the Hudson, the neglect that has crept over much of New York State has turned the curving slanted fields lined with piled stones partly back to forest. We see the land going to seed before our eyes. Young trees root insistently on open ground. This is transition forest coming down from the Adirondacks — a mix of boreal pine and the hardwoods that grow on the lower slopes of mountains. As we walk in the woods we pass through the lemon-green light of second-growth birch and hickory where the forest floor still has the grassy meadowlike quality of the sheep pasture it was a hundred years ago. The ground is strewn with chunks of white quartzite once assembled for walls, now scattered and sided with thick green moss like fur. Around them grows the autumn crop of gem-studded puffballs, the spangled panther, cinnabar chanterelles, pristine destroying angels, and huge russulas with sticky mottled heads. We follow the thin slapping of the falls into the deep cooling blue shade of the hemlocks. The floor here is matted with soft flakes of hemlock needles amid the broken bodies of fallen trees, their moist, splintering, deeply rotted wood stained with saffron streaks of witches' butter — the rot out of which new hemlocks grow.

Above the falls is what my great-aunt would once reverently have called the "drowned land," where the spills of an abandoned beaver colony unevenly divide a succession of small marshy lakes, lined with stumps gnawed into pencil points. The ghosts of trees stand knee-deep in the sludge of tussock grass and reeds — paper birch and sugar maple, carved by birds and the wind into the astonishingly beautiful shapes of ancient things. High hollowed outcroppings of limestone stand between the woods and water and we bring our picnics here sometimes, even in the snow. A broken landscape, the perfect territory for the animal we heard last night.

The eastern coyote is one of the mysteries of natural history at the turn of the twenty-first century, a time of massive die-off, of the rapid extinction of one species after another. This new animal has thrived in what might otherwise be called the destruction of habitat: the cutting back of old-growth forests; the creation of tarmac roads that run amid a patchwork of urban sprawl, strips of thinned woodland, field lots, and scrub. The animal is so discreet it fades away into brush and trees so that people are hardly aware of its presence. Its given name is part of its disguise — for everyone knows that a coyote is a harmless little animal that rarely exceeds thirty pounds, an animal of the desert and the grass plains, feeding on berries and lizards and carrion.

The eastern coyote hunts in packs, following herds of ungulates, as wolves, in their great variety, have always done. The wolf is a creature of the Ice Age. It came into being in the Pleistocene, when massive sheets of ice slid like razors around the world, cutting down the trees. The receding ice scraped and cleared the ground, creating tundralike expanses that became rich pastureland for great herds of roaming animals and the species that developed to feed on them: wolves and early man.

One might say that we are in a new ice age, the ice now concrete, metal, and asphalt — lifeless things, pushing the forest back, leaving only edges of insubstantial new growth; and that wolves, hunted almost to extinction a hundred years ago, have come back among us in this new form. A reemergence suggesting the spiral of insistent growth — destruction and decay followed by the blossoming of forms — that brings to mind the origin of animal in *anemos,* the Greek word for wind.

≥⊷

I used to think about a place of silver light like the negative of a photograph: Avalon, apple grove, the rotating island visible only to the dying; or the forest of precious stones that was the home of Tara, star, the wish-fulfilling wheel; or the island of Circe, circle, the white sea hawk with its circling flight. The silver forest was a realm of ghosts, a mirror world — like language, the great mirror that contains the reflections of everything that ever lived.

Language mirrors the fluidity of life itself. In its essence it is like molecular variation. There are patterns and root elements, and any hardened meaning is a kind of crusting over formed by use. Form is the Latin word for beauty, *forma.* The mind craves meaning as the eye craves beauty. They are essentially different levels of sense perception of the same thing, like variations on a chord of music. What the mind and the eye crave is form, spun out on a hidden underlying pattern, as the apple and the wolf in their succeeding forms are spun out on a patterned genetic underlay. One often finds both the visual beauty and the deeper meaning worked together in a single word. A vivid metaphorical conflation will often contain astute observations of natural history.

For years I have carried in my mind a fragment that is said to be the last thing that Catullus wrote. When I was young I was charmed by the intricacy of this small and perfect image formed of words — as though Catullus managed to capture a beautiful little animal on the page. His death is a mystery and people sometimes look to the fragment for clues. He had recently been devastated by the loss of his brother. Was his own death a suicide? The fragment suggests that he suffered some critical betrayal:

> Num te leaena montibus Libystinis
> Aut Scylla latrans infima inguinum parte
> Tam dura mente procreavit ac taetra
> In novissimo casu contemptam haberas
> O nimis fero corde

> A lioness in the Libyan mountains
> Or barking Scylla must have given birth to you
> With a mind so tainted and unmoved
> That you would have contempt for me
> In the ultimate crisis of my life
> Oh you with an iron heart

The lioness in the Libyan mountains is easy to track. This is a conventional insult. Among the desert people of the Mediterranean world, to whom the lion was once a reality, *ibn labwa*, son of a lioness, is far worse than "son of a bitch," for a lioness is a promiscuous animal — that lowest of all possible creatures, a female after sex. Scylla, on the other hand, opens a window through which one has a glimpse of something remarkable. Scylla was the sea monster that Circe warns Odysseus he will meet. But for Catullus, who used words with the skill of a jewelmaker, shading their colloquial, mythic, and historical sense together at once, there would be some hidden atmosphere or meaning. Scylla is often depicted as a woman who is "dogs from the waist down." This is actually just a linguistic pun on *skulax,* the Greek word for puppy. The Latin translation of *skulax* is — Catullus. Hence one might ask: is this some kind of joke? Is Catullus the one whose mother is a sea monster? But far more interesting is the epithet *latrans,* barking, for now we are on the solid ground of natural history. We are talking about something real.

Circe's story of a monster with many arms pulling sailors off the decks of ships and feeding them into a hidden mouth, bite by bite, is an accurate description of *architeuthis,* the giant squid. The Danish naturalist Steenstrup in 1857 wrote of "a fact not generally recorded in scientific literature, namely the wailing cries which [the squid] uttered . . . such sounds being likened to grunts, wails, moans, squeaks, miserable cries or yelps" heard over the ocean from miles away. Could this be the origin of the mermaid?

What I noticed about the giant squid in the American Museum of Natural History when I last saw it was that its sharp, birdlike beak was embedded in a hidden mouth the shape of a perfect octagon, textured like the seedbed of a lotus. The animal was red, faintly speckled over with pink polka dots, but I have read that waves of color pass like electrical currents over its skin. It has the largest living eye.

❧

We used to follow the route of an old Indian footpath that ran directly from our house on Seneca Lake across the state through the Mohawk Valley to my grandparents' house in Albany, where a huge old gingko tree stood in the back beyond a fishpond. We drove in the winter months through heavy snows to get there, through the black slate road cuttings of Utica when they glistened with ice, and Albany, the white city, rose before us in the distance above the white, frozen river.

My grandfather was a heavy man, as such men used to be,

with iron gray hair parted neatly in the middle in the old-fashioned way above his broad strong face. For forty years he was chief counsel and deputy commissioner of education for New York State. During my childhood he argued the landmark cases for censorship and for school prayer before the United States Supreme Court, and lost.

In my grandfather's office in the old state education department were books bound in dark green cloth that seemed to shine as though highly polished, the titles pressed into the spine of each volume in gold. These were the monographs on the natural history of the state that the department had commissioned for over a century.

What is New York State? Each of the green volumes gave detailed answers: a landscape created by water, the flood plain of the Great Lakes, the drainage patterns of great rivers, the migration routes — plant, bird, animal, and human — that belong to them.

The books presented New York State as a coherent entity, an aesthetic that has stayed with me all my life.

A bird illustrator at the state museum painted a picture for my grandfather in the 1940s that hung in our attic for years: a watercolor of the flowering branch of an apple tree. Among the blossoms was a bird of the richest purest blue.

One summer my husband and I drove east across America. Every night lightning shimmered through the dark and fell to the ground, starting brush fires in the dry grass all across Oregon, Washington, Idaho, Montana. When we stopped we found the diners and motels filled with yellow-suited firefighters streaked with black ash.

One night we slept on the Yellowstone River, at the base of hills covered with scorched, still-standing pines. In the morning we got up and drove for eighteen hours. Halfway across North Dakota in the dark we crossed a line. We could feel a dampness in the air — the smell of damp ground and high grass wet with dew and cornfields. In the dark I felt the great relief of the familiar presence of water. This was home, a territory within the far reach of the Great Lakes. In the morning we saw the low gray sky.

New York was called the Empire State because the waterways

that formed it — the Hudson, the St. Lawrence, the Mohawk, the Susquehanna — like human veins, rapidly enabled the penetration of the American continent.

When the eye studies a map of the eastern coast of North America, taking in the larger patterns, ignoring straight lines and boundaries and what they signify, what it sees is an arc — not a small accidental arc stranded in the sea, but part of a great formation, a ragged archipelago spanning the northern seas. Under the water from Ireland to Newfoundland along this curve, vast submarine plateaus run back to back across the Atlantic, forming a ghostly underwater chain that mirrors the land above. A gradual interaction developed over years and was part of a greater continuum, a topographical one. Water was the primary element in what became one of the most rapid transformations of land and culture in the history of the world. The sea brought fishermen from the northern coasts of Europe to the Grand Banks (where they brought back the fish that fed Catholic Europe for centuries). The fishermen made camps on the shore, where they dried the fish in the summer months to take back to Europe. The Indians came down to the shore, willing to trade even the clothes on their own bodies for the iron they saw in the hands of the fishermen, which instantly outmoded all of their painstakingly crafted implements of bark and bone and clay. The Europeans followed the sources of Indian fur inland, decimating as they went the beaver meadows along the waterways.

What became New York State in the nexus of these waterways was like a frame frozen in the process. Fort Orange at Albany was the dominant fur-trading outpost of the Dutch. The first wave of wealth built fortunes on beaver felt. The earliest state monograph, in 1844, reported that the beaver had entirely vanished from the state, which retained it as a state symbol denoting the creation of a financial center built on the back of the fur trade. In the decades that followed, the question became how to recognize and document what was native, how to bring the beaver back.

A 1918 study of New York native plants began:

One who is upon the gray ocean at this season of the year, when, in the woods and at the roadsides in the State of New York, the wild flowers are beginning to redeem their promises of life, appreciates as never before how much these quiet, persistent pioneers of the fields contribute in scent, color and form to the making of that which is summed up in the name New York . . . The sight of a spray of these native flowers, such as many a page in this book carries, would be as a twig borne back in ancient times to the ark — a sign that, though the flood of war has overwhelmed many valleys, the elemental processes of life go forward undisturbed in the "Empire State." (John Finley, New York State Museum Memoir 15, *Wild Flowers of New York*, 1918)

The long progression of water plants, from cattails and sedges to pickerelweed and butter yellow spatterdock, showed how the primitive spiraling form of the water lily contained the gradual development of the flower — as though the practical use of beauty was thought out in a slow deliberate effort by the organism of the plant itself.

The water lily affords the best illustration of the evolution of blossoms we have in the world today. In the several species of the family we may see sepals being transformed into petals, and petals into stamens; the transition is so gradual and insensible that many intermediate bodies are neither petals nor stamens, but partly both.

The plates that follow display every subtle gradation of blue in the color spectrum, from the soft blue grays of wild flags to the deep indigo of the closed gentian — its blossoms sealed like lips — and the pale violet-tinged cerulean of the roadside bluebell with its tough fibrous stems and leaves. The scarlet lobelia, the cardinal flower, was already a rare sight in the woods, and people were told to stop picking them before they disappeared.

The Fishes of New York (1903) contains a peculiar photograph of a boy straddling an immense dead sturgeon. The sturgeon was so common in the Hudson at the time that it was called Albany beef and formed the basis of a caviar industry that eclipsed the caviar industry of Russia. The fish was a living fossil from the Devonian, six to twelve feet long, and often breached:

> This is a species concerning which so many stories have been related as to leaping into boats and injuring the occupants.

Of its eyes:

> the pupils are black; the iris golden.

Of the books in my grandfather's office, the one I prized the most was Memoir 12, the two-volume Birds of New York, by Elon Howard Eaton, illustrated by Louis Agassiz Fuertes. For years I kept odd fragments of Eaton's text like pieces of beautiful ribbon tucked away in books:

> Here the kingfisher forms her nest of fish bones and scales ejected from her stomach.

> The Jack or Hudsonian curlew: like the long-billed curlew it exhibits much sympathy for wounded companions, often sacrificing its life by returning in answer to their cries. Its flesh is much inferior to that of the Eskimo curlew, being quite unpalatable, except in the fall when it has fed for some time on berries and grasshoppers.

> The willet, woodcock (scolopax), phalarope, turnstone, plover, snipe or sandpiper are all variations on the same theme, whether they resemble pebbles or leaves.

The observations recorded in the books from not fifty years before astonished us: of white pelicans on Seneca Lake; of a winter rookery on East Lake Road in 1910 that was estimated to contain thirty thousand crows; of a golden eagle trapped in Rochester and taken to the zoo; of the "Niagara trap" — the placid waters of the Niagara River that every year swept hundreds of unsuspecting whistling swans over the falls; of the meticulously recorded disappearance of the passenger pigeon, which only a few decades before still darkened the sky with a flapping of wings that sounded like thunder or the noise of the ocean or an approaching train. The birds came down in the millions at the salt licks of the Montezuma Swamp on the Seneca River.

Eaton taught at Hobart. Fuertes lived in Ithaca. The two men marked the north-south boundaries of the Finger Lakes, and their detailed local observations centered the books there. They began with water birds — not sea birds, but the freshwater birds of the Great Lakes.

The heavy plates of the illustrations were carefully faced in opaque sheets of tissue paper, as though they were something precious that needed to be wrapped to keep from fading away.

Louis Agassiz Fuertes was named after the Swiss naturalist Agassiz, whose view that there was an underlying pattern in nature, and that the pattern and its inherent beauty were a manifestation of the divine, was rendered obsolete by Darwin. Agassiz discovered the Ice Age and used the layering of the exposed rock face of the Helderberg escarpment outside of Albany to establish its chronology. He developed his expertise as a naturalist and paleontologist with his work on the fossil fish of the Devonian. When Agassiz was teaching at Harvard and a young man would come to his office to ask if he could apply to be his student, Agassiz would set before him a fish preserved in formaldehyde. He would leave the young man alone there for hours with the instruction, "Look at your fish."

❧

The New York State Museum on the fifth floor of the old state education building was one of the joys of my childhood. The elevator doors would open, and there you were in the Gilboa Forest, the Devonian forest of giant fern trees found in the fossil river delta of heaped mud that formed the Catskill Mountains. It was the oldest remnant forest in the world.

The Devonian was the age of rising up, away from the water, the age of bones and stems. In the Devonian the moon was half the distance to the earth that it is today. The tides, drawn up by the closeness of the moon, pulled the earth faster in space. The four hundred days of the Devonian year were recorded in the carbonate skins of horn corals formed in the shallow tropical sea of the Finger Lakes, whose evaporates would ultimately become a salt formation hundreds of feet thick. The bodies of once living things trapped in layers of mud became the oil pools of Cuba Springs south of Seneca Lake. The drilling equipment from the salt mines was used to drill the first simple oil wells when petroleum began to replace whale oil during the blockades of the Civil War. Petroleum, like whale oil, is fat, fossil fat — the decayed, transformed fat of living things trapped in mud and clay, oozing forth with either a green or reddish cast, depending upon whether its host matter was animal or plant.

The discovery of petroleum in the United States is credited to the Jesuit D'Allion, who in the early seventeenth century saw the Seneca use as medicine the black material that bubbled up from the ground in western New York.

Sullivan's men had similar reports and in 1750 the French officer Contrecoeur wrote to Montcalm of the Seneca ceremonially setting fire to the oil on the surface of the Allegheny River.

John D. Rockefeller grew up two lakes east of Seneca, on Owasco Lake. His grandfather had been a snake oil ("Seneca Oil") salesman out of Richford south of Ithaca.

🐝

The first sound you heard coming into the Gilboa Forest was water. The central feature of the reconstruction was a waterfall, like the falls at Slate Rock, Taughannock, or Watkins Glen — the water pooling beautifully down through layers of mud stone, broken symmetrically into layered blocks. Among the blocks were the ossified trunks of the giant fern trees.

In glass-faced dioramas on either side were the giant celadon dragonflies of the Devonian, and the first amphibians coming out on the shore with their bare softcolored skin, some speckled, some scarlet. These were the first delicate land animals, which would become huge carnivores and shrink again into living flowers.

Behind the waterfall were glass cases in the walls showing the brilliantly colored depths of the Devonian sea — the ammonites, with their flowing colored bodies, straight and curled in spiral shells; a shard of the shell of a giant squid found in the earth near Watertown; the living trilobites, whose fossil bodies would click out of the blue-gray stone on our cottage shore in perfect ovals, finely lined as though in carved relief. Then the early fish, whose fins ultimately became hands; and the Albany rhipidistrian, with scales delicately outlined in red, whose body explained the development of lungs — a stomachlike sac to hold air during the dry periods when the water was gone, just as the vascular insides of rising plants developed, when blue-green algae from the ancient seas washed exposed surfaces of dry rock, to hold moisture away from the water.

The first fragments of life were scraps of color, pieces of pigment that by way of the color itself could draw in and transform light into life. Thus color and physical form were intimately bound, form deriving from color. The color turned the light into sugar, the substance of life.

I sometimes feel I am living the life of my great-aunt Dorothy, who lived with her husband, a World War I vet, and her dog in a house in the woods in the lake country of southern Ontario. I can still hear her old Scots voice in my head:

Notice the ends of their wings — whether they're jagged or smooth. They fly around like that for miles till one of them finds a carcass,

*and they all come for the feast. I think they're turkey vultures,
'cause that's typical of turkey vultures. There were a bunch of them
eating a dead skin when I was on my way to church one day. They
waited till we were oh, twenty feet away or so and then they took
off, slow as you please, and sat in a plane tree nearby. Oh, they're
horrible-looking things, with their red necks. They nest in just a pile
of rocks on the ground.*

*That little lake was so clear when we moved here. Now it's all weeds
from one end to the other. Likely the pollution in the water is killing
it. All these hideous hydro-things. They annihilate life.*
*   "I don't know about that, Aunt Dorothy," Mother says. "I don't
want to live without electricity. I'm too old and set in my ways."*
*   Oh, I lived that way for years. I don't mind it at all. We loved the
lamps. Ern and I.*

*There's still part of an Indian family around here. Dug up long be-
fore we ever came to Verona. They took out human remains, human bones. I don't think that there're any skeletons. It's more likely
to have been a skirmish.*

*The bare trees — they're just like filigree. Where Marg lived there
are thick woods and they see wolves or coyotes. I don't know what
they are. I could drive all day, just looking at rocks and trees. Are
you still gathering rocks? Remember how we used to have to stop so
you could pick up rocks? Look at all the spots on that rock. Lichen,
is it?*

When I graduated from college I was visiting Aunt Dorothy in
her little house on Mud Lake. I asked her what she thought I should
do. I wanted to go to graduate school to become an archaeologist. "If
I were you, I'd take the money and make a down payment on a farm
somewhere in New York State," she said.

Blue-green algae from the ancient seas washed exposed surfaces of dry rock and ultimately thickened into moss, the reproduction of which involves swimming particles that do not need to leave the watery element of the plant. In another three hundred million years woody plants would grow so far from their origin in water, and from each other as they rose up into the air that their reproduction would need to depend on something else, another kind of living being and the means to attract it. In the Cretaceous, as the flower developed and diversified in scent and color and form, populations of insects developed and diversified across the earth. The first known honeybee dates to a fossil from the Cretaceous found in Kazakhstan.

✿

My husband once gave me a painting of a white farmhouse beside a maple tree. Beyond its red barns and fields a ridge of blue hills rose through a green wash of forest. It was a familiar local scene that could have been anywhere in the open land of the hidden hill country of New York State, in the Hudson Valley or the Finger Lakes.

When I first came to the valley where we live now, I thought how like the painting it was. It was a quiet afternoon in early May and the apple trees in the orchard behind the house were blossomed over with white.

I walked up through the fields to the line of pines above and found an abandoned cistern, its sides matted with soft green moss. Water trickled in from a spring cut through a face of layered mud stone, swirled with silver and rust that crumbled at a touch.

The water was filled with silky masses of eggs. Newts hung loosely down below the surface. A bull-headed salamander ate one as I watched, tearing the flesh away as it held the newt firmly in its hands. Mottled black mink frogs with bright green masks watched as I slid into the cold water and swam.

As I lifted myself up on the side of cracked old cement, its ragged surface stinging my hands, I heard, then felt, the wind come across the valley. A blotch of white unfolded from the bare trees and circled down on the wind, a white hawk.

When we moved into the farmhouse, it had not been lived in for years. Inside we found only a leather-top card table with lime-green legs, and on it five large clay pots of red geraniums. At first we heard strange sounds in the night — footsteps like those of an old man pacing a slow familiar round, getting up to make coffee in the early morning dark, a ghost. One night we heard the rocking chair creaking back and forth in the next room. Lan walked warily in with his flashlight, and uttered a cry of delight. Curled on the narrow back of the chair was a flying squirrel watching him with huge round eyes as bright as black onyx. Folds of fur as soft as velvet were tucked along its sides. The little squirrel examined us in a placid, almost benevolent way. It hopped about our bedroom, climbing up to watch us in bed — where I sat drinking hot whiskey as Lan read *Bleak House* out loud to put us back to sleep. At last the squirrel climbed into a geranium by the open window, twitching its whiskers, clusters of soft white, like dandelion fluff. Out it ran. Then back it came to look again.

The year is a reliable circle: the bluebirds nesting in the apple trees in spring, the spiraling song of the veery in summer, the swarms of dragonflies hunting midges over the fields in the damp evenings of early fall, the black bear that comes down from the woods before the snow and smashes our beehives to pieces, neatly scraping off each comb and tossing them all in a broad white sweep up the hill. Every year something is a little different and plucks at our attention.

It often seems that the things I looked for in the days of Bob and Gary, almost as though they were mysteries, are all around me now. The wild turkeys that began to appear on East Lake Road years ago are thirty strong in our field. I woke one day to see one walking down the road in the melting snow. As he turned into the sun I was amazed to see that the feathers puffed out over his breast and body had a bright metallic green sheen. His sky blue head and red wattle made him the most extraordinary apparition. The females scratching through the snow for seeds ignored him, however.

Once I walked along the edge of the field with our dog Fred, a slender red hound with velvet-soft ears. Someone once said that Fred was the dog of the Magyar hordes of Central Asia — poor companionable Fred, nudging me along through my daily routines with his little head. He has a habit of racing around me as I walk, running around and around in a figure eight. On the final loop he likes to run right toward your legs and just miss them, as a joke — though once in a while he will crash and knock you down. Then he picks up a heavy fallen branch and, growling, shakes and drags it along the ground, as he once might have dragged a wild boar.

I have seen a large old buck in the field locking antlers with younger ones that come to throw him. They dance back and forth, the tines of their antlers curled together, until the old one, who is slightly crippled in a back leg, throws the younger to the ground.

One winter Fred and I saw a wounded doe lying crumpled in the field. As we came near her the buck with the bad back leg bounded out of the trees below us and ran to the doe and nudged her to her feet with his antlers. She stumbled and fell. He nudged her to her feet again, and again, slowly helping her reach the safety of the far trees.

I have heard people say that valleys like this were made by beaver, by the water from the beaver meadows eating the rock. The work of the animal is everywhere in evidence. Beaver are said to be responsible for all the meadows in the woods, but it is their continuous creation of wet land, of the source of life, that is the most remarkable of their achievements. In a wetland one sees life threaded together in a vivid interwoven tapestry. Fish rise to swallow midges through the transparent surface film. Dragonflies dash over the surface after the midges, clicking their wings with a soft papery sound, and are abruptly taken by the fish. A heron stands watching in a shallow lip of water like a stripped discolored branch.

Inside the beaver dam, I have read, the floor is covered with sweet rushes and pine boughs, and above the water level there is a hidden balcony for the beaver to look out from. In the beaver marshes above the falls there once grew rushes and sedges with their diamond-shaped three-sided stems and knotty beaded heads, and brightly colored irises, wild flags — violet blue in spring. We would often see deer, stepping slowly through the marsh among the water plants, unaware of our presence. Birds filled the place and its rotting trees. One fall evening as I walked there with Fred on the path of dead leaves a woodcock suddenly flew up from the ground, looking like the dead leaves themselves, whirred around and around us, making its vibrating cry that sounds like wax paper over a comb — *peet peet.*

Then the beaver left and the marsh began to dry up and to turn into a meadow. We do not often go there anymore.

Lan and Fred and I walked up the hill today, looking for the white hawk, which we have not seen all week. The trees, still bare silver in the smoky light of early spring, brighten the air down the valley as the sun slants through.

We did not see the hawk, and looked at the ice that still covered the cistern in white-veined edging around the middle as it began to crack, melting into a pool where the spring runs in.

We walked along the upper edge of the field and into the woods where a path leads along a stone wall into stands of old birch. There are many fallen trees, and those still standing are heavily pocked by pileated woodpeckers. We walked through stands of tall white pine, their lower branches broken, their faint fine green needles in a lime blue wash of sap spilling down from above.

We walked down the hill toward the creek. There was a tongue of ice, like spilled water, long and broad, but frozen, coming into the scrub growth. Lan and I went over to see what it was, and followed it toward its source among the osier thickets. To our surprise we saw the ice broaden into the lakelike bed of a marsh, drowning the huge old trees that were girdled about by intricate gnawing, though still standing, others fallen leaving only sharp new points. In their midst, through the gray stems of osiers in the water laced with ice, we saw a high dam of newly packed mud. It had a quality of fairyland. The beaver must have created it in just a few months, for the last time we had walked down to the lower fields there was only a narrow stream flowing under the trees. We walked down a deer path through the dry reeds and saw that this is where the white hawk had gone. It perched in a dead tree above the marsh, keenly eyeing its wealth of living things.

As we came down to the water two wood ducks flew up with shrill whistling cries. A beaver appeared in the water below us. His large black tail floated up behind him as he stood, like a little man, half out of the water, picking up clumps of waterweeds and eating them with his hands. We sat watching, with our tea and bread and thick crystalline honey, for half an hour or more, the light thickening on the far birches and the slope of trees.

The beaver at last swam away, and we began to walk back toward the house. Then we saw him surface again, and stand in a shallow pool directly in our path. Fred ran, barking, to the pool's edge. The beaver slapped the water loudly with his heavy tail. The sound resounded all around us, sending circles through the water, through the air. Then he disappeared. I looked to the slope of trees, which was turning a pinkish gold. We walked away into the woods, the trees like golden threads before us.